I0494228

Disclaimer

Book Title: Smoke Component Yields from Bench-scale Fire Tests: 2. ISO 19700 Controlled Equivalence Ratio Tube Furnace

Book Author: Nathan D. Marsh; Richard G. Gann; Marc R. Nyden;

Book Abstract: A standard procedure is needed for obtaining smoke toxic potency data for use in fire hazard and risk analyses. Room fire testing of finished products is impractical, directing attention to the use of apparatus that can obtain the needed data quickly and at affordable cost. This report presents examination of the second of a series bench-scale fire tests to produce data on the yields of toxic products in both pre-flashover and post-flashover flaming fires. The apparatus is the ISO/TS 19700 controlled equivalence ratio tube furnace. This apparatus uses a mechanical feed mechanism to supply solid fuel into a tube furnace at a pre-determined rate, so that the global equivalence ratio can be adjusted. The test specimens were cut from finished products that were also burned in room-scale tests: a sofa made of upholstered cushions on a steel frame, particleboard bookcases with a laminated finish, and household electric cable. Initially, the standard test procedure was followed for two fire stages, well ventilated flaming and post- flashover. Subsequent variation in the procedure included dicing the specimen, further decreasing the equivalence ratio (well ventilated flaming) or increasing it (post-flashover), increasing the mass loading while maintaining the equivalence ratio, and increasing the fuel feed rate while maintaining the equivalence ratio.

The yields of CO_2 CO, HCl, and HCN were determined. The yields of other toxicants (NO, NO_2, formaldehyde, and acrolein) were below the detection limits, but volume fractions at the detection limits were shown to be of limited toxicological importance relative to the detected toxicants. In general, the largest effects were seen between the two fire stages. The other variations within the fire stage had minor effects on gas yields. Under post-flashover conditions, the sum of the CO_2 and CO yields frequently accounted for half or less of the carbon originally in the specimen. As a result, the gaseous combustion products cannot be used to estimate the mass burning rate. Under post flashover conditions, the CO yield for the sofa approached the value of 0.2 found in real-scale postflashover fire tests. However, for the bookcase and cable it did not. Yields of HCl from the cables generally approached their notional yields under well- ventilated conditions, and HCN was most readily detected from the sofa under post-flashover conditions at toxicologically significant concentrations.

Citation: NIST TN - 1761

Keyword: fire; fire research; smoke; room fire tests; fire toxicity; smoke toxicity

NIST Technical Note 1761

Smoke Component Yields from Bench-scale Fire Tests: 2. ISO 19700 Controlled Equivalence Ratio Tube Furnace

Nathan D. Marsh
Richard G. Gann
Marc R. Nyden

http://dx.doi.org/10.6028/NIST.TN.1761

National Institute of
Standards and Technology
U.S. Department of Commerce

NIST Technical Note 1761

Smoke Component Yields from Bench-scale Fire Tests: 2. ISO 19700 Controlled Equivalence Ratio Tube Furnace

Nathan D. Marsh
Richard G. Gann
Marc R. Nyden
Fire Research Division
Engineering Laboratory

http://dx.doi.org/10.6028/NIST.TN. 1761

December 2013

U.S. Department of Commerce
Penny Pritzker, Secretary

National Institute of Standards and Technology
Patrick D. Gallagher, Under Secretary of Commerce for Standards and Technology and Director

National Institute of Standards and Technology Technical Note 1761
Natl. Inst. Stand. Technol. Tech. Note 1761, 45 pages (December 2013)
http://dx.doi.org/10.6028/NIST.TN. 1761
CODEN: NTNOEF

ABSTRACT

A standard procedure is needed for obtaining smoke toxic potency data for use in fire hazard and risk analyses. Room fire testing of finished products is impractical, directing attention to the use of apparatus that can obtain the needed data quickly and at affordable cost. This report presents examination of the second of a series bench-scale fire tests to produce data on the yields of toxic products in both pre-flashover and post-flashover flaming fires. The apparatus is the ISO/TS 19700 controlled equivalence ratio tube furnace. This apparatus uses a mechanical feed mechanism to supply solid fuel into a tube furnace at a pre-determined rate, so that the global equivalence ratio can be adjusted. The test specimens were cut from finished products that were also burned in room-scale tests: a sofa made of upholstered cushions on a steel frame, particleboard bookcases with a laminated finish, and household electric cable. Initially, the standard test procedure was followed for two fire stages, well ventilated flaming and post-flashover. Subsequent variation in the procedure included dicing the specimen, further decreasing the equivalence ratio (well ventilated flaming) or increasing it (post-flashover), increasing the mass loading while maintaining the equivalence ratio, and increasing the fuel feed rate while maintaining the equivalence ratio.

The yields of CO_2 CO, HCl, and HCN were determined. The yields of other toxicants (NO, NO_2, formaldehyde, and acrolein) were below the detection limits, but volume fractions at the detection limits were shown to be of limited toxicological importance relative to the detected toxicants. In general, the largest effects were seen between the two fire stages. The other variations within the fire stage had minor effects on gas yields. Under post-flashover conditions, the sum of the CO_2 and CO yields frequently accounted for half or less of the carbon originally in the specimen. As a result, the gaseous combustion products cannot be used to estimate the mass burning rate. Under post flashover conditions, the CO yield for the sofa approached the value of 0.2 found in real-scale postflashover fire tests. However, for the bookcase and cable it did not. Yields of HCl from the cables generally approached their notional yields under well-ventilated conditions, and HCN was most readily detected from the sofa under post-flashover conditions at toxicologically significant concentrations.

Keywords: fire, fire research, smoke, room fire tests, fire toxicity, smoke toxicity

This page intentionally left blank

TABLE OF CONTENTS

LIST OF FIGURES

LIST OF TABLES

I. INTRODUCTION

A. CONTEXT OF THE RESEARCH

Estimation of the times that building occupants will have to escape, find a place of refuge, or survive in place in the event of a fire is a principal component in the fire hazard or risk assessment of a facility. An accurate assessment enables public officials and facility owners to provide a selected or mandated degree of fire safety with confidence. Without this confidence, regulators and/or designers tend to apply large safety factors to lengthen the tenable time. This can increase the cost in the form of additional fire protection measures and can eliminate the consideration of otherwise desirable facility designs and construction products. Error in the other direction is also risky, in that if the time estimates are incorrectly long, the consequences of a fire could be unexpectedly high.

Such fire safety assessments now rely on some form of computation that takes into account multiple, diverse factors, including the facility design, the capabilities of the occupants, the potential growth rate of a design fire, the spread rates of the heat and smoke, and the impact of the fire effluent (toxic gases, aerosols, and heat) on people who are in or moving through the fire vicinity.[1] The toolkit for these assessments, while still evolving, has achieved some degree of maturity and quality. The kit includes such tools as:

- Computer models of the movement and distribution of fire effluent throughout a facility.

  - Zone models, such as CFAST[2], have been in use for over two decades. This model takes little computational time, a benefit achieved by simplifying the air space in each room into two zones. A number of laboratory programs, validation studies,[3] and reconstructions of actual fires have given credence to the predictions.[4]

  - Computational fluid dynamics (CFD) models, such as the Fire Dynamics Simulator (FDS)[5], have seen increased use over the past decade. FDS is more computationally intense than CFAST in order to provide three-dimensional temperature and species concentration profiles. There has been extensive verification and validation of FDS predictions.[5]

  These models calculate the temperatures and combustion product concentrations as the fire develops. These profiles can be used for estimating when a person would die or be incapacitated, *i.e.*, is no longer able to effect his/her own escape.

- Devices such as the cone calorimeter[6] and larger scale apparatus[7], which are routinely used to generate information on the rate of heat release as a commercial product burns.

- A number of standards from ISO TC92 SC3 that provide support for the generation and use of fire effluent information in fire hazard and risk analyses.[8] Of particular importance is ISO 13571, which provides consensus equations for estimating the human incapacitating exposures to the narcotic gases, irritant gases, heat, and smoke generated in fires.[9]

More problematic are the sources of data for the production of the harmful products of combustion. Different materials can generate fire effluent with a wide range of toxic potencies. Most furnishing and interior finish products are composed of multiple materials assembled in a variety of geometries, and there is as of yet no methodology for predicting the evolved products from these complex assemblies. Furthermore, the generation of carbon monoxide (CO), the most common toxicant, can vary by orders of magnitudes, depending on the fire conditions.[10]

An analysis of the U.S. fire fatality data[11] showed that post-flashover fires comprise the leading scenarios for life loss from smoke inhalation. Thus, it is most important to obtain data regarding the generation of harmful species under post-flashover (or otherwise underventilated) combustion conditions. Data for pre-flashover (well-ventilated) conditions have value for ascertaining the importance of prolonged exposure to "ordinary" fire effluent and to short exposures to effluent of high potency.

B. OBTAINING INPUT DATA

The universal metric for the generation of a toxic species from a burning specimen is the yield of that gas, defined as the mass of the species generated divided by the consumed mass of the specimen.[12] If both the mass of the test specimen and the mass of the evolved species are measured continuously during a test, then it is possible to obtain the yields of the evolved species as the burning process, and any chemical change within the specimen, proceeds. If continuous measurements are not possible, there is still value in obtaining a yield for each species integrated over the burning history of the test specimen.

The concentrations of the gases (resulting from the yields and the prevalent dilution air) are combined using the equations in ISO 13571 for a base set of the most prevalent toxic species. Additional species may be needed to account for the toxic potency of the fire-generated environment.

To obtain an indicator of whether the base list of toxic species needs to be enhanced, living organisms should also be exposed to the fire effluent. The effluent exposure that generates an effect on the organisms is compared to the effect of exposure to mixtures of the principal toxic gases. Disagreement between the effluent exposure and the mixed gas exposure is an indicator of effluent components not included in the mixed gas data or the existence of synergisms or antagonisms among the effluent components. This procedure has been standardized, based on data developed using laboratory rats.[13,14] However, it is recognized that animal testing is not always possible. In these cases, it is important to identify, from the material degradation chemistry, a reasonable list of the degradation and combustion products that might be harmful to people.

Typically, the overall effluent from a harmful fire is determined by the large combustibles, such as a bed or a row of auditorium seats. The ideal fire test specimen for obtaining the yields of effluent components is the complete combustible item, with the test being conducted in an enclosure of appropriate size. Unfortunately, reliance on real-scale testing of commercial products is impractical, both for its expense per test and for the vast number of commercial products used in buildings. Such testing *is* practical for forensic investigations in which there is knowledge of the specific items that combusted.

A more feasible approach for obtaining toxic gas yields for facility design involves the use of a physical fire model – a small-scale combustor that captures the essence of the combustible and of the burning environment of interest. The test specimen is an appropriate cutting from the full combustible. To have confidence in the accuracy of the effluent yields from this physical fire model, it must be demonstrated:

- How to obtain, from the full combustible, a representative cutting that can be accommodated and burned in the physical fire model;

- That the combustion conditions in the combustor (with the test specimen in place) are related to the combustion conditions in the fire of interest, generally pre-flashover flaming (well ventilated or underventilated), post-flashover flaming, pyrolysis, or smoldering;

- How well, for a diverse set of combustible items, the yields from the small-scale combustor relate to the yields from real-scale burning of the full combustible items; and

- How sensitive the effluent yields are to the combustor conditions and to the manner in which the test specimen was obtained from the actual combustible item.

At some point, there will be sufficient data to imbue confidence that testing of further combustibles in a particular physical fire model will generate yields of effluent components with a consistent degree of accuracy.

The National Institute of Standards and Technology (NIST) has completed a project to establish a technically sound protocol for assessing the accuracy of bench-scale device(s) for use in generating fire effluent yield data for fire hazard and risk evaluation. In this protocol, the yields of harmful effluent components are determined for the real-scale burning of complete finished products during both pre-flashover and post-flashover conditions. Specimens cut from these products are then burned in various types of bench-scale combustors using their standard test protocols. The test protocols are then varied within the range of the combustion conditions related to these fire stages to determine the sensitivity of the test results to the test conditions and to provide a basis for improving the degree of agreement with the yields from the room-scale tests.

This report continues with a brief description of the previously conducted room fire tests. The full details can be found in Reference 15. Following this recap are the details of the tests using the second of four bench-scale apparatus to be examined.

C. PRIOR ROOM-SCALE TESTS

1. Test Configuration

With additional support from the Fire Protection Research Foundation, NIST staff conducted a series of room-fire tests of three complex products.[15] The burn room was 2.44 m wide, 2.44 m high, and 3.66 m long (8 ft x 8 ft x 12 ft). The attached corridor was a 9.75 m (32 ft) long extension of the burn room. A doorway 0.76 m (30 in.) wide and 2.0 m (80 in.) high was centered in the common wall. The downstream end of the corridor was fully open.

2. Combustibles

Four fuels were selected for diversity of physical form, combustion behavior, and the nature and yields of toxicants produced. Supplies of each of the test fuels were stored for future use in bench-scale test method assessment.

- "Sofas" made of up to 14 upholstered cushions supported by a steel frame. The cushions consisted of a zippered cotton-polyester fabric over a block of a flexible polyurethane (FPU) foam. The fire retardant in the cushion padding contains chlorine atoms. Thus, this fuel would be a source of CO_2, CO, HCN, HCl, and partially combusted organics. The ignition source was the California TB133 propane ignition burner[16] faced downward, centered over the center of the row of seat cushions. In all but two of the tests, the sofa was centered along the rear wall of the burn room facing the doorway. In the other two tests, the sofa was placed in the middle of the room facing away from the doorway to compare the burning behavior under different air flow conditions. Two of the first group of sofa tests were conducted in a closed room to examine the effect of vitiation on fire effluent generation. In these, an electric "match" was used to initiate the fires.

- Particleboard (ground wood with a urea formaldehyde binder) bookcases with a laminated polyvinylchloride (PVC) finish. This fuel would be a source of CO_2, CO, partially combusted organics, HCN, and HCl. To sustain burning, two bookcases were placed in a "V" formation, with the TB133 burner facing upward under the lower shelves.

- Household wiring cable, consisting of two 14 gauge copper conductors insulated with nylon, an uninsulated ground conductor, two paper filler strips, and an outer jacket of plasticized PVC. This fuel would be a source of CO_2, CO, HCl, and partially combusted organics. Two cable racks containing 3 trays each supported approximately 30 kg of cable in each of the bottom two trays and approximately 17 kg in each of the middle and top trays. The cable trays were placed parallel to the rear of the burn room. Twin propane ignition burners were centered under the bottom tray of each rack.

The elemental chemistry of each combustible was determined by an independent testing laboratory. More details regarding the elemental analysis can be found in NIST TN 1760[17]. The elemental composition of the component materials in the fuels is shown in Table 1.

Table 1. Elemental Analysis of Fuel Components.

Sample	C	H	N	Cl	P	O
Bookcase	0.481 ± 0.6 %	0.062 ± 0.8 %	0.029 ± 13 %	0.0030 ± 4 %	NA	0.426 ± 1 %
Sofa	0.545 ± 1 %	0.080 ± 1 %	0.100 ± 1 %	0.0068 ± 16 %	0.0015 ± 17 %	0.267 ± 4 %
Cable	0.576 ± 0.5 %	0.080 ± 1.5 %	0.021 ± 6 %	0.323 ± 0.4 %	NA	NA

The uncertainties are the standard deviation of three elemental analysis tests, combined with the uncertainty of the mass fraction of individual material components, i.e. fabric vs. foam for the sofa.

D. PHYSICAL FIRE MODELS

Historically, there have been numerous bench-scale devices that were intended for measuring the components of the combustion effluent.[18,19] The combustion conditions and test specimen configuration in the devices vary widely, and some devices have flexibility in setting those conditions. Currently, ISO TC92 SC3 (Fire Threat to People and the Environment) is proceeding toward standardization of one of these devices, a tube furnace (ISO/TS 19700[20]) and is considering standardization of another, the cone calorimeter (ISO 5660-1[21]) with a controlled combustion environment. There are concurrent efforts in Europe and ISO to upgrade the chemical analytical capability for a closed box test (ISO 5659-2[22]). Thus, before too long there may well be diverse (and perhaps conflicting) data on fire effluent component yields available for any given product. This situation does not support either assured fire safety or marketplace stability.

Detailed characterizations of the ISO/TS 19700 tube furnace were first published by Blomqvist *et al.*[23] and Stec *et al.*[24] At the same time a comparison was conducted between this apparatus and the ISO 9705 room scale tests.[25] More recently, the tube furnace has been used to study the effect on smoke particle-size distribution[26] and smoke and hydrocarbon yields[27] of different fire retardant strategies.

The apparatus used for this work was constructed according to the 2006 version of ISO/TS19700. That Technical Specification has since been updated.

This page intentionally left blank

II. EXPERIMENTAL INFORMATION

A. SUMMARY OF ISO 19700 APPARATUS

1. Hardware

The ISO/TS 19700 tube furnace consists of three main parts: (1) a quartz tube running through an electrically heated furnace; (2) a 30 L dilution and sampling chamber; and (3) a specimen boat and drive mechanism, which can advance the specimen into the furnace at a controlled rate. Aspects of this apparatus are depicted in Figure 1 through Figure 3. Air is supplied at both the upstream end of the quartz tube and in the dilution and sampling chamber. By controlling the upstream air flow rate and the specimen feed rate, the equivalence ratio in the tube furnace can be adjusted to model several fire stages. In this work, we focus on two stages—well ventilated flaming and postflashover fires, and set the temperature and equivalence ratio as specified in ISO/TS 19700.

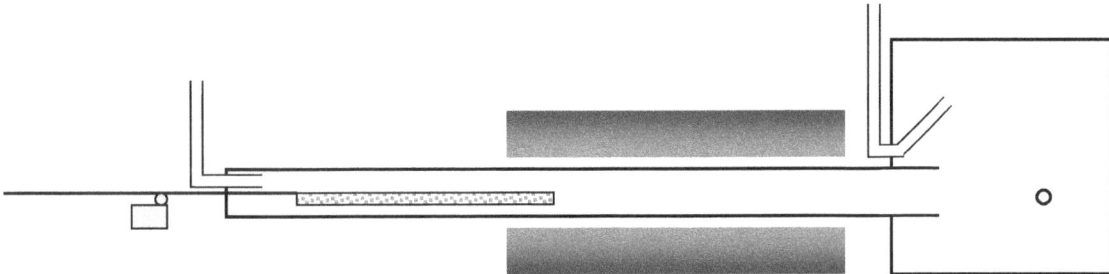

Figure 1. Schematic of the ISO/TS 19700 Tube Furnace

Figure 2. Photograph of the NIST ISO/TS 19700 Tube Furnace

Figure 3. Closeup of the Sampling and Dilution Chamber

The furnace temperature is controlled by a proportional–integral–derivative (PID) control module, which reads the temperature from a type-K thermocouple located 1 cm above the quartz tube in the center of the furnace. This means that the setpoint for the furnace will actually control the maximum heat flux imposed on the specimen—at the ends of the furnace the flux will be considerably lower. The Technical Specification calls for the points where the temperature falls 100 ºC below the maximum temperature be located between 125 mm and 250 mm from the location of the maximum temperature. Temperature profiles of the NIST furnace are shown in Figure 4 and Figure 5. These profiles were recorded with gas flow rates consistent with those used during tests. However, it is important to note that thermocouples receive radiant energy from the heating elements, while heat transfer to the transparent gas is less effective; therefore the gas temperature in the tube furnace could be considerably lower.

The issue is further complicated when a specimen is burning in the tube. Our furnace has an active control system which can compensate somewhat for the additional heat input, but this is not a requirement of the Technical Specification. In any event, the combustion effluent is hot, and the gas temperature will be higher when it is present. It may also be optically thick, in which case it can be heated radiatively by the heating elements.

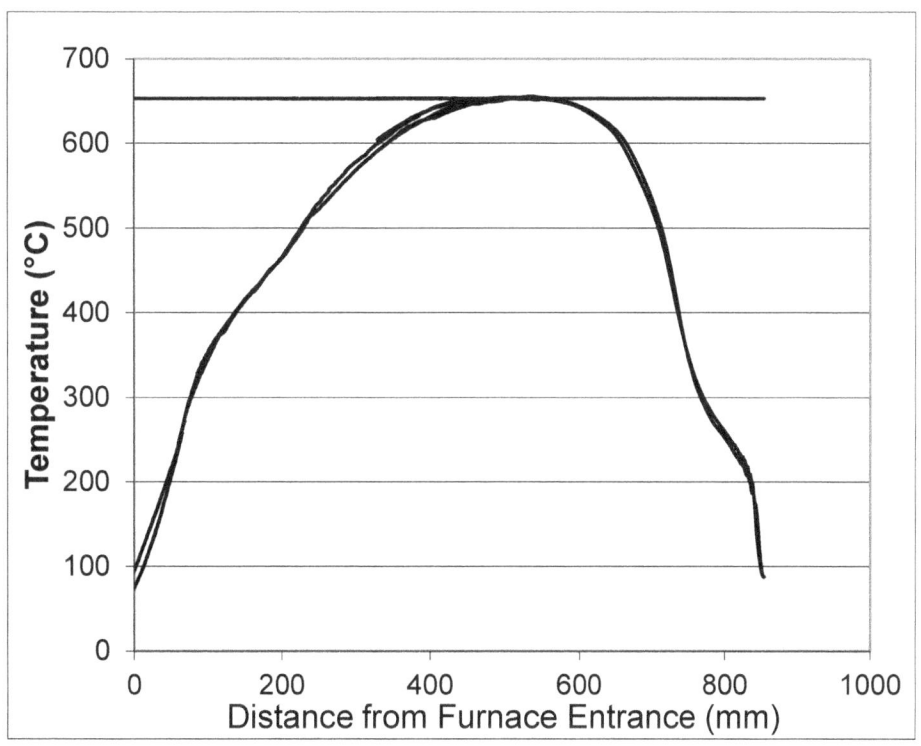

Figure 4. Furnace Temperature Profile at a 650 ℃ Setpoint

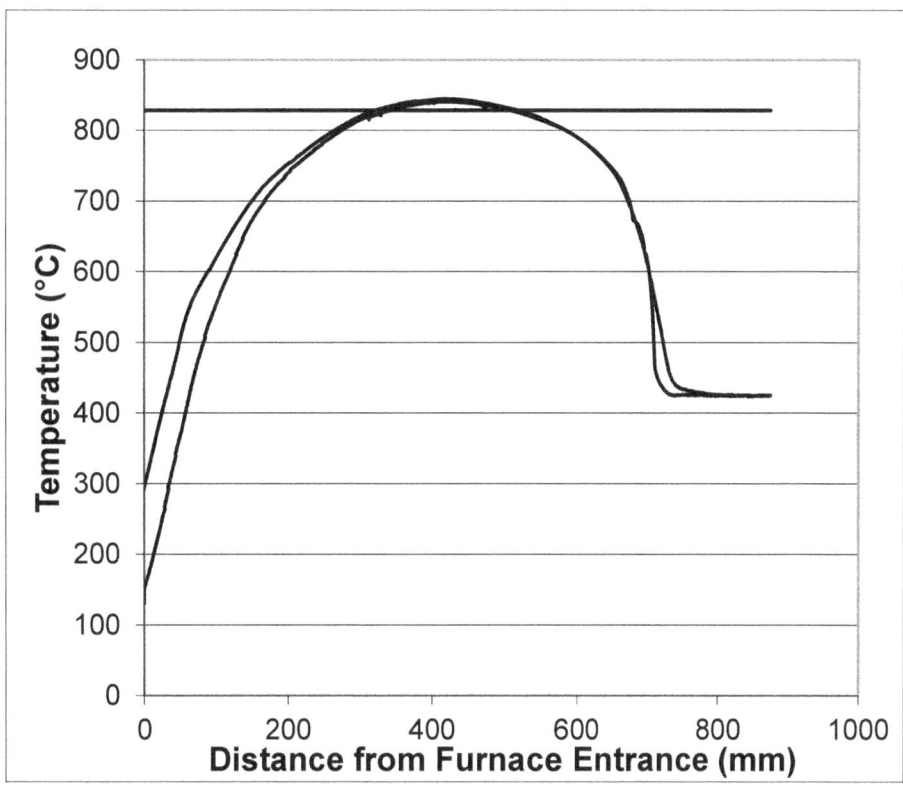

Figure 5. Furnace Temperature Profile at a 825 ℃ Setpoint

2. Test Operation

Test specimens are placed in the sample boat, which is then placed in the (cool) upstream end of the quartz tube. Once the appropriate temperature and air flow rate are established, data collection begins and the drive mechanism is activated to advance the specimen into the furnace. At the same time, a portion of the exhaust from the dilution chamber is diverted to gas analysis instruments. Ideally, the combustion of the specimen reaches some kind of steady state, from which gas concentrations can be established and yields derived. In our experience, complex items like multi-conductor cables burn in a more fluctuating and periodic way than, for example, simple thermoplastics. However, it is still possible to identify ignition and extinction events, and to determine average yields from the period in between.

The actual equivalence ratio, which can differ from the one intended if the fraction of combustible in the specimen varies, was calculated based on oxygen depletion:

$$\Psi_O = \frac{D_{O2} \times 1330}{C_{m.loss}}$$

$$\phi = \frac{\dot{m}_{loss} \times \Psi_O}{O}$$

where D_{O2} is the oxygen depletion in the mixing and measurement chamber, as a volume fraction; $C_{m.loss}$ is the mass-loss concentration of the test specimen, in grams per cubic meter ($g \cdot m^{-3}$); m_{loss} is the mass-loss rate of the test specimen, in milligrams per minute ($mg \cdot min^{-1}$); Ψ_O is the stoichiometric oxygen mass to fuel mass ratio; O is the oxygen supply rate, in milligrams per minute ($mg \cdot min^{-1}$), given by the following equation:

$$O = P \times 0.209\ 5 \times 1\ 330,$$

where P is the primary air flow rate, in liters per min ($l \cdot min^{-1}$); 1 330 is a factor to convert the volume of oxygen to mass of oxygen at 20 °C.

The oxygen depletion is also used to calculate the necessary primary air flow for an equivalence ratio of 2.0 using the formula $P = D_{O2} \times 1.193$. This assumes that all of the material consumed in well-ventilated conditions will be consumed in underventilated conditions as well.

Testing at each set of conditions was performed in triplicate. We report the concentrations measured in the dilution chamber, averaged over a period of time between 4 min and 12 min from the beginning of the experiment. (Although steady state burning, as defined in the current version of ISO/TS 19700 was not observed for our specimens, we were able to identify this time interval as falling between ignition- and extinction-related phenomena, and found that selecting data from different parts of this interval did not significantly alter the calculated average. However, given the characteristic mixing time of the dilution chamber of 60 s, at least 200 s of data should be used for calculating an average concentration.) The three time-averaged concentrations are then averaged. The reported uncertainty includes the standard deviation of this final average, which represents the run-to-run variation in the experiment.

3. Gas Sampling and Analysis Systems

In the room-scale tests (Section I.C), measurements were made of 12 gases. Water and methane were included because of their potential interference with the quantification of the toxic gases. Two of the toxic gases, HBr and HF, were not found in the combustion products because there was no fluorine or bromine in the test specimens. The remaining eight toxic gases were acrolein (C_3H_4O), Carbon monoxide (CO), carbon dioxide (CO_2), formaldehyde (CH_2O), hydrogen chloride (HCl), hydrogen cyanide (HCN), nitric oxide (NO), and nitrogen dioxide (NO_2). Some of these turned out to be generated at levels that would not have contributed significantly to the incapacitation of exposed people. Thus, it was deemed unlikely that animal tests would have added much tenability information. As a result, the same gases were monitored in the bench-scale tests, and no animals were exposed. The basis for comparison between tests of the same combustibles at the two scales is the yields of the chemically diverse set of toxicants.

CO and CO_2 were quantified using a nondispersive infrared (NDIR) gas analyzer; oxygen was quantified by a paramagnetic analyzer in the same instrument. The precisions of the analyzers, as provided by the manufacturer, were:

CO: 10 µL/L

CO_2: 0.02 L/L

O_2: 0.05 L/L

For sampling from the dilution and sampling, gas was continuously drawn from through copper tubing of 6.35 mm outer diameter. The flow passed through a coiled tube immersed in a water ice bath; an impinger bottle immersed in dry ice, with its upper half filled with glass wool; and finally a glass fiber disk filter before reaching the pump and then finally the analyzer. The traps and filter removed particulates and condensable species, including water, that would otherwise interfere with and possibly harm the analyzer. While sampling, the flow was maintained at 1 L/min for the CO and CO_2 detectors and 0.2 L/min for the O_2 detector. The analyzer itself was calibrated daily with zero and span gases (a mixture of 5000 µL/L CO and 0.08 L/L of CO_2 in nitrogen, and ambient air (0.2095 L/L oxygen on a dry basis)). The span gas is certified to be accurate to within 2 % of the value.

The concentrations of CO and the additional six toxic gases were measured using a Midac Fourier transform infrared (FTIR) spectrometer[*] equipped with an stainless steel flow cell (2 mm thick ZnSe windows and a 0.1 m optical pathlength), maintained at (170 ±5) °C. Samples were drawn through a heated 6.35 mm (¼ in.) stainless steel tube from the dilution chamber. The sample was pulled through the sampling line and flow cell by a small pump located downstream from the flow cell. There were no traps or filters in this sampling line. The pump flow was measured at 4 L/min maximum, but was at times lower due to fouling of the sampling lines with smoke deposits. The instrument could collect data at 1 Hz and could also combine several scans per recorded spectrum. There did not appear to be any advantage to either approach, other than that recording fewer spectra required less storage.

An example of a spectrum measured by FTIR spectroscopy during one such test is displayed in Figure 6. The series of peaks extending from about 3050 cm^{-1} to 2600 cm^{-1} are due to HCl. In this case, it is possible to resolve the individual frequencies corresponding to changes in the population of rotational states as the H-Cl bonds vibrate. This is usually only possible for small gas phase molecules. There are three spectral features due to CO_2 that are evident in this

---

[*] Certain commercial equipment, products, or materials are identified in this document in order to describe a procedure or concept adequately or to trace the history of the procedures and practices used. Such identification is not intended to imply recommendation, endorsement, or implication that the products, materials or equipment are necessarily the best available for the purpose.

spectrum. The most intense, centered at 2350 cm⁻¹, corresponds to asymmetric stretching of the two C=O bonds. The symmetric stretch is not observed because there is no change in dipole moment when both O atoms move in phase. The second feature, seen as two distinct peaks centered at about 3650 cm⁻¹, is an overtone band that derives from the simultaneous excitation of these bond-stretching modes. The third peak at about 650 cm⁻¹ is due to the out of plane bending of the molecule. There are bands due to the C≡O stretching vibrations in carbon monoxide, centered at about 2150 cm⁻¹. The remaining peaks in this spectrum are due to H_2O.

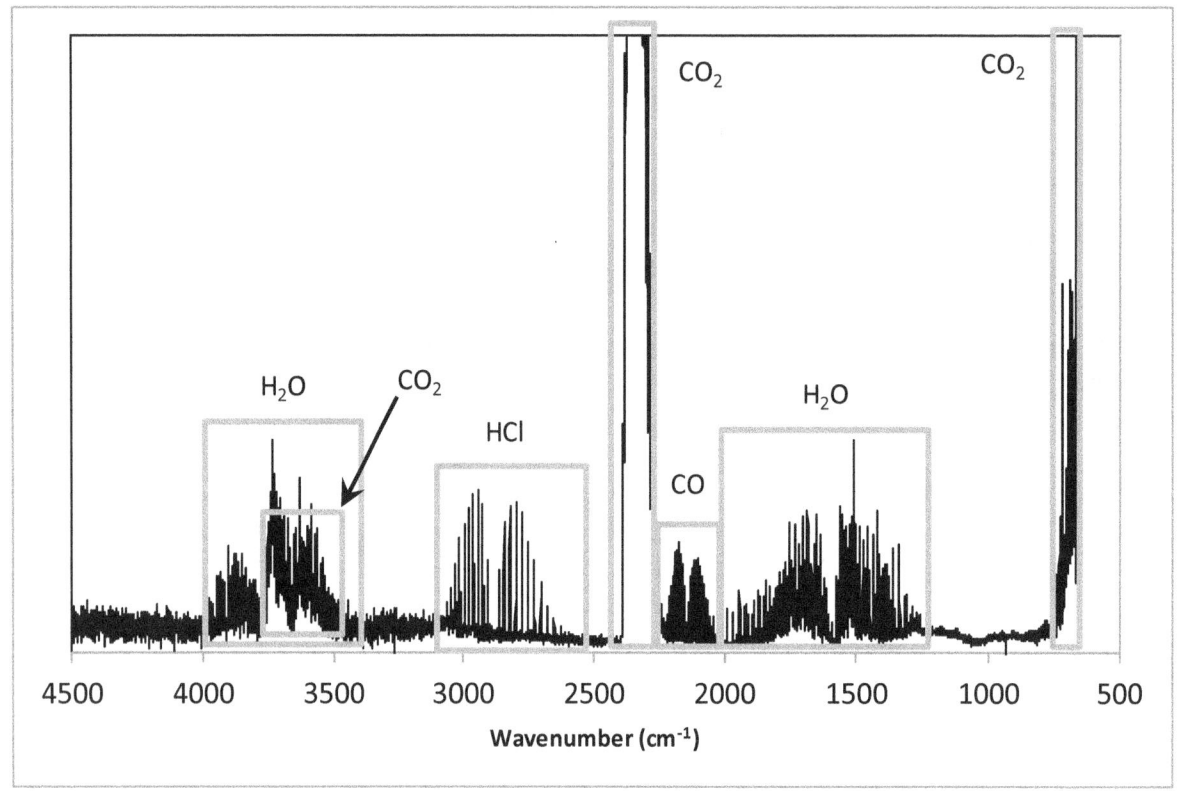

Figure 6. FTIR Spectrum of the Products of Burning Electrical Cable

Using these spectra, gas concentrations were quantified using the Autoquant software. This is a software package for performing real time and off-line quantitative analyses of target compounds, and is based on the Classical Least Squares (CLS) algorithm as described by Haaland et al.[28] In this method, the measured spectra are fit to linear combinations of reference spectra corresponding to the target compounds.

Calibration spectra were obtained from a quantitative spectral library assembled by Midac[29] and from a collection of spectra provided the Federal Aviation Administration who performed bench-scale fire tests on similar materials.[30] In this analysis, the least squares fits were restricted to characteristic frequency regions or windows for each compound that were selected in such a way as to maximize the discrimination of the compounds of interest from other components present in fire gases. All reference spectra were recorded at 170 °C and ambient pressure.

The identities of the target compounds (as well as other compounds that absorb at the same frequencies and must, therefore, be included in the analyses), their corresponding concentrations (expressed in units of µL/L for a mixture of the calibration gas and N_2 in a 0.1 meter cell), and the characteristic spectral windows used in the quantitative analyses are listed in Table 2.

Also listed in this Table are minimum detection limits (MDLs) for each of the target compounds. These values, which represent the lowest concentrations that can be measured with the instrumentation employed in these tests, were estimated as follows. The calibration spectra were added to test spectra (which, when possible, were selected in such a way that only the compound of interest was not present) with varying coefficients until the characteristic peaks of the target compounds were just discernible above the baseline noise. The value of signal averaging over *ca.* 100 spectra was included. The MDL values reported in Table 2 were obtained by multiplying these coefficients by the known concentrations of the target compound in the calibration mixtures.

Water, methane and acetylene are included in the quantitative analyses because they have spectral features that interfere with the target compounds. The nitrogen oxides absorb in the middle of the water band that extends from about 1200 cm^{-1} to 2050 cm^{-1}. Consequently, the real limits of detection for these two compounds are an order of magnitude higher than for any of the other target compounds. Thus, it is not surprising that their presence was not detected in any of the tests.

Table 2. Species and Frequency Windows for FTIR Analysis.

Compound	Reference Volume Fraction (µL/L)	Frequency Window (cm⁻¹)	Minimum Detection Limit (µL/L)
CH_4	483	2800 to 3215	20
C_3H_4O	2250	850 to 1200	20
CH_2O	11300	2725 to 3000	40
CO_2	47,850	660 to 725, 2230 to 2300	800[a]
CO	2410	2050 to 2225	20
H_2O	100,000	1225 to 2050, 3400 to 4000	130[a]
HCl	9870	2600 to 3100	20
HCN	507	710 to 722, 3200 to 3310	35
NO	512	1870 to 1950	70
NO_2	70	1550 to 1620	40

[a] Present in the background.

Delay times for gas flows from the sampling locations within the test structure to the gas analyzers were small compared to the duration of the specimen burning. The burn durations were near 20 min for all specimens. Combining the gas sample pumping rate and the volumes of the sampling lines, the delay times were about 5 s for both the fixed-gas and FTIR instruments. These delay times are enough to allow for a small degree of axial diffusion. However, since our analysis integrates the data over time, this did not adversely affect the quantification of total gas evolved.

B. OPERATING PROCEDURES

1. "Standard" Testing

The intent was to test specimens of each of the three types under condition simulating well-ventilated and post-flashover conditions. The steps in the procedure are:

- Calibrate the fixed-gas analyzer using zero and span gases, and for the oxygen span, ambient air.

- Establish furnace temperature and gas flow rates for the chosen condition.

- Weigh the specimen

- Begin data acquisition from the gas analyzers.

- Open the upstream end of the quartz tube.

- Load the sample boat containing the specimen.

- Close the quartz tube.

- Activate the feed mechanism.

- Observe the ignition and burning of the specimen

- Once the feed mechanism has withdrawn the specimen, quench any residual combustion with nitrogen.

- Weigh the sample boat containing any residue.

2. Test Specimens

Specimens were prepared by cutting a 450 mm length of one of the three specimen types, with a cross section selected to give a combustible mass loading of approximately 25 mg·mm^{-1}. This resulted in bookcase specimens that were nominally 7 mm square in cross section, sofa foam specimens that were nominally 25 mm square in cross section, and cable specimens that were single lengths cut from the spool. Each sofa specimen was covered with a strip of the upholstery fabric that was 25 mm by 450 mm. All specimens were kept in a conditioning room for at least 24 hours before an experiment. Specimens were weighed before and after each experiment so that the total mass loss, and therefore the average mass loss rate, could be determined.

3. Test Procedure Variation

One of the purposes of this program was to obtain effluent composition data in tests with variants on the standard operating procedure. This would enable examination of the potential for an improved relationship with the yield data from the room-scale tests, as well as an indication of the sensitivity of the gas yields to the specified operating conditions.

- Variation in specimen conformation: specimens were tested as intact, i.e. as cut from a larger item and preserving the item's structure as much as possible, and "diced", i.e. cut up into 5 mm pieces and, in the case of the cable, separating the constituent layers.

- Changing the air flow rate, increasing by 30 % for the well ventilated condition and decreasing by 30% for the postflashover condition.

- Doubling the mass loading by adding an additional length of specimen, but maintaining the equivalence ratio by increasing the primary air flow rate accordingly.

- Since the sofa specimens were already dimensionally as large as would fit in the tube furnace, they were cut in half instead, with the primary air flow altered to maintain the equivalence ratio.

- Increasing the feed rate by 50 %, but maintaining the equivalence ratio by increasing the primary air flow accordingly.

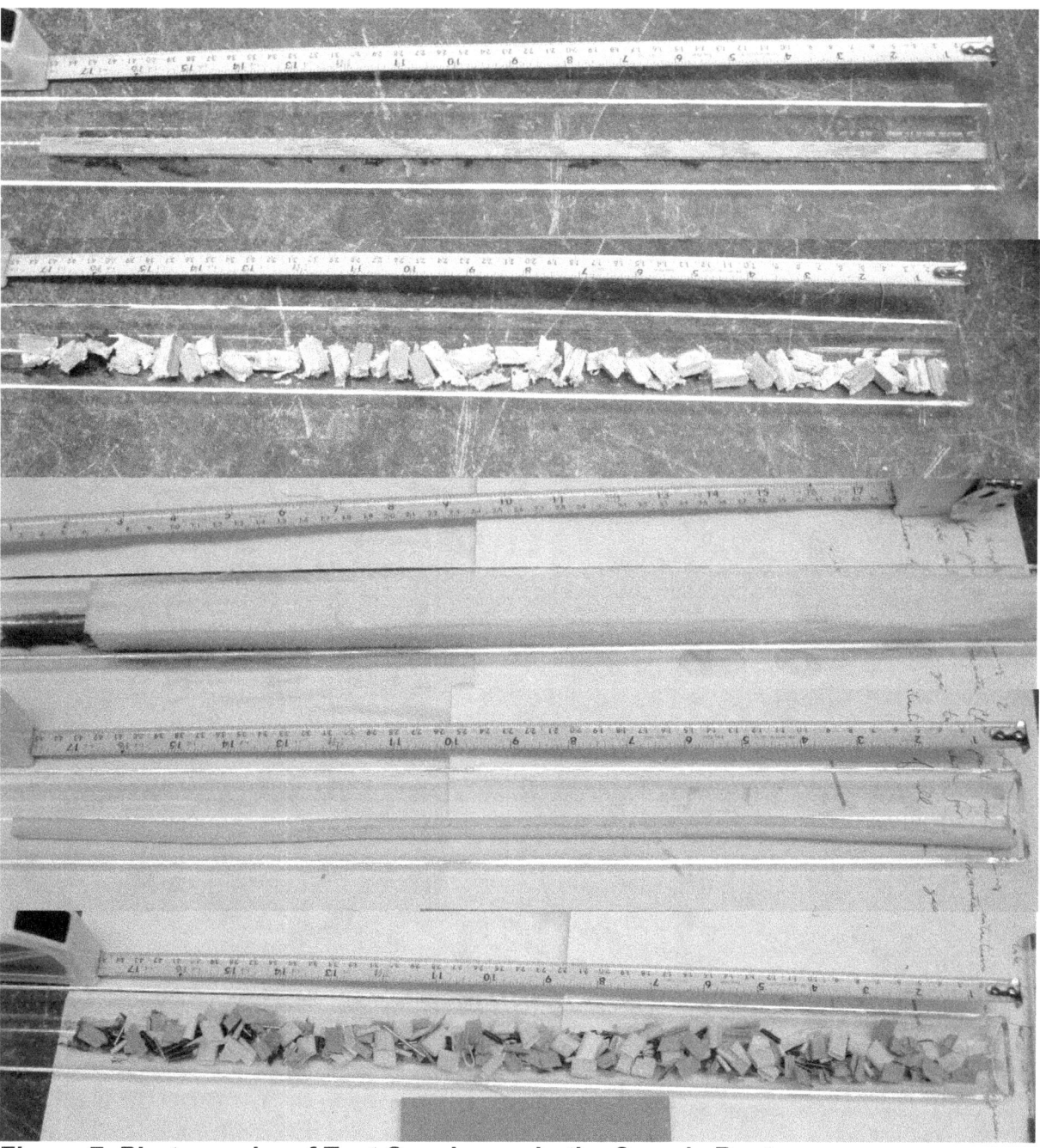

Figure 7. Photographs of Test Specimens in the Sample Boat

C. DATA COLLECTION

The data from the fixed gas analyzer were recorded on a personal computer using a custom-made data acquisition system based on National Instruments data acquisition hardware. Values were recorded at 1 s intervals. The FTIR spectra were recorded using the software package provided by the manufacturer. Spectra were recorded every 1 s to 6 s.

III. CALCULATION METHODS

A. MASS LOSS RATE

The specimen mass loss during a test was determined from the initial and final mass of the specimen in the sample boat, which has an uncertainty of 0.1 g. The mass loss rate is calculated from the linear fuel load and the linear feed rate. This assumes a steady state is quickly reached. Mass loss rates may actually be higher at the beginning and end of the experiment, implying a lower mass loss rate than calculated for the main portion of the experiment. This could result in underestimation of gas yields.

B. NOTIONAL GAS YIELDS

The notional, or maximum possible, gas yields (Table 3) were calculated as follows:

- CO_2: Assume all the carbon in the test specimen is converted to CO_2. Multiply the mass fraction of C in the test specimen (
- Table 1) by the ratio of the molecular mass of CO_2 to the atomic mass of carbon.
- CO: Assume all the carbon in the test specimen is converted to CO. Multiply the mass fraction of C by the ratio of the molecular mass of CO to the atomic mass of carbon.
- HCN: Assume all the nitrogen in the test specimen is converted to HCN. Multiply the mass fraction of N by the ratio of the molecular mass of HCN to the atomic mass of nitrogen.
- HCl: Assume all the chlorine in the test specimen is converted to HCl. Multiply the mass fraction of Cl by the ratio of the molecular mass of HCl to the atomic mass of chlorine.

The notional yields from the bookcase and cable specimens were assumed to be the same as the yields from the intact combustibles.[15] The sofa specimen had a mass ratio of fabric to foam that differed modestly from the intact sofas.

Table 3. Calculated Notional Yields of Toxic Products from the Test Specimens.

Gas	Notional Yields		
	Bookcase	**Cable**	**Sofa**
CO_2	1.72 ± 1 %	2.11 ± 1 %	1.95 ± 4 %
CO	1.09 ± 1 %	1.33 ± 1 %	1.24 ± 4 %
HCN	0.057 ± 13 %	0.040 ± 6 %	0.193 ± 4 %
HCl	0.0026 ± 4 %	0.332 ± 1 %	0.0069 ± 19 %

The uncertainty in the notional yield values is determined by the uncertainty in the prevalence of the central element (in the bullets just above) in the combustible. For the cuttings from the sofas, the uncertainty in the notional yields was increased by the small variability, (estimated at 3 percent) in the relative masses of the fabric and padding materials in the test specimen.

C. CALCULATED GAS YIELDS

1. CO and CO_2

Yields of CO and CO_2 were calculated using gas concentrations from the NDIR instrument averaged over several minutes, the measured flow rates of primary and secondary air, the temperature in the dilution chamber, and mass loss rate.

$$Y = (M/V_m) \times (F_v / C_{m.loss}) \times 10$$

Where M is the molar mass of the gas (g mol^{-1})

V_m is the molar volume at STP i.e. 24.055 mol L^{-1}

F_v is the volume fraction of the gas, in %

$C_{m.loss}$ is the mass-loss concentration in g m^{-3}, given by:

$$C_{m.loss} = m_{loss} \times feed\ rate\ /\ volume\ flow\ rate$$

Where m_{loss} is Δm / specimen length (mg mm^{-1}), the feed rate is in mm min^{-1}, and the volumetric flow rate is in L min^{-1}, (corrected by the ideal gas law for increased temperature.)

As we observed previously[17], the CO_2 absorption band in the FTIR is saturated at normal volume fractions and is therefore highly non-linear. As both the FTIR and NDIR measurements were taken from the same location, it was not necessary to measure the CO_2 concentration by FTIR spectroscopy.

2. HCl and HCN

The only calculable HCN yields were from the sofa specimens. The only calculable HCl yields were from the cable specimens. The FTIR spectra from these experiments were analyzed as described in section II.A.2. This analysis normally includes an uncertainty of 10 % of the reported value. Volume fractions were converted into yields using the formula above.

3. Other Gases

The volume fractions of the other toxic gases were always below the detection limit. Thus, the upper limits of the yields of these gases were estimated using their limits of detection.

IV. RESULTS

A. TESTS PERFORMED

The following is the test numbering key, with format F-C-T-P-R-L-N, where

- F: Fuel [S = sofas; B = bookcases; C = cable]
- C: 1 for intact, 2 for diced
- T: Furnace temperature, 650 for well ventilated or 825 for post-flashover
- P   Primary air flow rate (rounded to nearest L min^{-1})
- R: Sample boat feed rate in mm min^{-1}
- L: Fuel load, 1 for normal, 2 for double, 5 for one-half
- N: Replicate test number for that set of combustible and conditions

Run codes ending with a letter "a" or "b" indicate that the run was repeated because of an error.

Table 4 through Table 9 present the test data and the calculated yields for the bookcase, sofa, and cable specimens, respectively. The equivalence ratio, Φ, is calculated from the mass loss rate, the stoichiometric oxygen to fuel mass ratio, and the oxygen supply rate from the primary air flow rate. (The stoichiometric oxygen to fuel mass ratio is calculated from the oxygen depletion, the mass loss rate, and the total air flow rate in the well-ventilated tests for a given specimen type.)

The horizontal shaded bands highlight groups of replicate tests.

Table 4. Data from Bookcase Material Tests

Code	Specimen Mass (g)	Mass Loss Rate (mg/min)	Primary Airflow (L/min)	Φ	NDIR F_{CO2} (L/L)	F_{CO} (μL/L)	D_{O2} (L/L)	FTIR F_{CO} (μL/L)	
b-1-650-10-**40**-1-001	14.8	1320	11.6	0.43	0.0168	< 10	0.019	30	
b-1-650-10-**40**-1-002	14.3	1220	11.6	0.42	0.0164	< 10	0.018	30	
b-1-650-10-**40**-1-003	14.8	1260	11.6	0.44	0.0171	< 10	0.019	40	
b-2-650-10-**40**-1-001	14.7	1310	11.6	0.43	0.0169	< 10	0.019	60	
b-2-650-10-**40**-1-002	13.7	1150	11.6	0.41	0.0162	< 10	0.018	100	
b-2-650-10-**40**-1-003	13.9	1160	11.6	0.35	0.0164	< 10	0.015	50	
b-1-650-13-**40**-1-001a	14.6	1230	15.1	0.34	0.0184	< 10	0.019	70	
b-1-650-13-**40**-1-002	14.2	1180	15.1	0.35	0.0181	< 10	0.020	70	
b-1-650-13-**40**-1-003	14.9	1260	15.1	0.38	0.0194	< 10	0.021	40	
b-1-650-20-**40**-2-001	28.1	2360	22.2	0.48	0.0359	< 10	0.039	30	
b-1-650-20-**40**-2-002	29.5	2460	23.1	0.46	0.0364	< 10	0.040	50	
b-1-650-20-**40**-2-003	31.4	2580	23.1	0.44	0.0347	< 10	0.038	< 20	
b-1-650-15-**60**-1-001	16.1	1910	17.3	0.37	0.0152	< 10	0.024	20	
b-1-650-15-**60**-1-002	14.3	1730	17.4	0.41	0.0114	< 10	0.026	30	
b-1-650-15-**60**-1-003	15.8	1840	17.4	0.49	0.0149	< 10	0.032	20	
b-1-825-2-**40**-1-001	15.5	1200	2.61	1.89	0.0057	1660	0.007	1420	
b-1-825-2-**40**-1-002	14.8	1180	2.57	1.89	0.0030	830	0.002	790	
b-1-825-2-**40**-1-003	15.8	1210	2.57	1.93	0.0032	780	0.004	920	
b-2-825-2-**40**-1-001	15.1	1170	2.57	1.87	0.0032	560	0.004	690	
b-2-825-2-**40**-1-002	13.5	1080	2.57	1.72	0.0030	620	0.004	< 20	
b-2-825-2-**40**-1-003	15.5	1220	2.57	1.94	0.0033	880	0.004	700	
b-1-825-1-**40**-1-001	15.6	1160	1.76	2.71	0.0026	1280	0.004	110	
b-1-825-1-**40**-1-002	14.8	1150	1.77	2.67	0.0023	830	0.003	600	
b-1-825-1-**40**-1-003	15.1	1200	1.77	2.77	0.0023	1210	0.003	1090	
b-1-825-4-**40**-2-001	26.9	2130	5.00	1.75	0.0067	1540	0.008	1760	
b-1-825-4-**40**-2-002	28.8	2230	5.00	1.83	0.0065	2180	0.007	1830	
b-1-825-4-**40**-2-003	29.1	2260	5.00	1.85	0.0059	1750	0.007	1690	
b-1-825-3-**60**-1-001	15.7	1830	3.53	2.12	0.0042	1380	0.005	1240	
b-1-825-3-**60**-1-002	13.8	1630	3.39	1.97	0.0044	1240	0.006	1090	
b-1-825-3-**60**-1-003	14.3	1630	3.38	1.97	0.0044	1060	0.005	1240	

Table 5. Data from Sofa Material Tests

Code	Specimen Mass (g)	Mass Loss Rate (mg/min)	Primary Air flow (L/min)	Φ	NDIR F_{CO_2} (L/L)	NDIR F_{CO} (μL/L)	NDIR D_{O_2} (L/L)	FTIR F_{CO} (μL/L)	FTIR F_{HCN} (μL/L)
s-1-650-10-**40**-1-001	10.3	870	11.6	0.44	0.0152	300	0.019	510	< 35
s-1-650-10-**40**-1-002	10.5	920	11.6	0.36	0.0141	350	0.016	490	< 35
s-1-650-10-**40**-1-003	10.2	840	11.6	0.46	0.0154	340	0.020	490	< 35
s-1-650-13-**40**-1-001	10.5	890	15.0	0.18	0.0074	190	0.010	530	< 35
s-1-650-13-**40**-1-002	10.8	940	15.1	0.18	0.0075	290	0.010	570	40
s-1-560-13-**40**-1-003	10.7	940	15.1	0.20	0.0082	440	0.011	710	60
s-1-650-5-**40**-5-001	7.8	680	6.0	0.21	0.0043	120	0.005	340	< 35
s-1-650-5-**40**-5-002	7.5	640	5.8	0.28	0.0044	100	0.006	460	< 35
s-1-650-5-**40**-5-003	7.8	680	5.8	0.26	0.0040	80	0.006	340	< 35
s-1-650-15-**60**-1-001	10.4	1290	17.3	0.22	0.0106	110	0.014	420	< 35
s-1-650-15-**60**-1-002	10.4	1320	17.4	0.19	0.0093	50	0.013	590	< 35
s-1-650-15-**60**-1-003	10.7	1350	17.4	0.17	0.0084	< 10	0.011	420	< 35
s-1-825-2-**40**-1-001	10.5	870	2.92	1.25	0.0043	1810	0.007	1680	90
s-1-825-2-**40**-1-002	13.7	920	2.92						
s-1-825-2-**40**-1-003	9.7	760	2.92	1.08	0.0042	1600	0.007	1460	80
s-1-825-1-**40**-1-001	10.4	910	2.01	1.89	0.0027	2320	0.005	2250	130
s-1-825-1-**40**-1-002	10.4	900	2.00	1.88	0.0030	2150	0.005	1880	120
s-1-825-1-**40**-1-003	10.2	850	2.18	1.64	0.0030	2050	0.005	1630	100
s-1-825-1-**40**-5-001a	7.4	600	1.18	2.14	0.0009	670	0.002	290	< 35
s-1-825-1-**40**-5-002a	7.0	520	1.18	1.83	0.0010	600	0.002	840	50
s-1-825-1-**40**-5-003a	6.1	510	1.65	1.29	0.0010	470	0.000	700	40
s-1-825-2-**60**-1-001	13.5	1200	1.88	2.67	0.0030	2610	0.005	2350	180
s-1-825-2-**60**-1-002b	11.0	1280	1.89	2.84	0.0036	2700	0.006	2260	190
s-1-825-2-**60**-1-003	10.4	1240	1.88	2.76	0.0035	2910	0.007		

Table 6. Data from Cable Material Tests

Code	Specimen Mass (g)	Mass Loss Rate (mg/min)	Primary Air flow (L/min)	Φ	NDIR F_{CO_2} (L/L)	F_{CO} (μL/L)	D_{O_2} (L/L)	FTIR F_{CO} (μL/L)	F_{HCl} (μL/L)
c-1-650-10-**40**-1-001	39.2	950	11.6	0.36	0.0108	990	0.016	1120	2760
c-1-650-10-**40**-1-002	39.3	640	11.6	0.36	0.0108	900	0.016	1020	2720
c-1-650-10-**40**-1-003	39.1	960	11.6	0.34	0.0099	1310	0.015	1200	2660
c-2-650-10-**40**-1-001	39.2	960	11.5	0.40	0.0120	1160	0.017	1190	3320
c-2-650-10-**40**-1-002	39.6	890	11.6	0.38	0.0112	1280	0.017	1260	3050
c-2-650-10-**40**-1-003a	39.4	970	11.6	0.37	0.0110	1260	0.016	1070	2890
c-1-650-13-**40**-1-001	39.1	950	15.1	0.26	0.0099	880	0.015	910	2520
c-1-650-13-**40**-1-002	38.7	920	14.3	0.29	0.0103	1030	0.015	950	2480
c-1-650-13-**40**-1-003	39.2	940	15.1	0.26	0.0098	920	0.015	930	2690
c-1-650-20-**40**-2-001	77.7	1810	22.8	0.32	0.0197	1900	0.027	1310	3610
c-1-650-20-**40**-2-002	79.0	1940	23.1	0.32	0.0205	1740	0.028	1270	3850
c-1-650-20-**40**-2-003	77.3	1920	23.1	0.30	0.0188	2050	0.026	1960	4470
c-1-650-15-**60**-1-001	38.6	1330	17.4	0.32	0.0144	1270	0.021	1190	3580
c-1-650-15-**60**-1-002	38.8	1350	17.4	0.33	0.0151	1280	0.022	1120	3710
c-1-650-15-**60**-1-003	38.6	1350	17.4	0.32	0.0147	1240	0.021	1110	3500
c-1-650-10-**20**-2-001	76.4	930	10.0	0.50	0.0145	1420	0.021	1590	3380
c-1-650-10-**20**-2-002	75.4	920	10.0	0.47	0.0134	1340	0.020	1580	3380
c-1-650-10-**20**-2-003	76.5	930	10.0	0.48	0.0139	1350	0.020	1620	3480
c-1-825-2-**40**-1-001	39.6	950	2.23	1.71	0.0026	500	0.004	670	1680
c-1-825-2-**40**-1-002	39.8	940	2.23	1.70	0.0032	670	0.005	570	1510
c-1-825-2-**40**-1-003	40.0	950	2.23	1.71	0.0030	610	0.005	460	1140
c-1-825-2-**40**-1-001	39.1	940	2.26	1.68	0.0029	570	0.003	700	1380
c-1-825-2-**40**-1-002	38.6	940	2.23	1.69	0.0024	480	0.004	760	1530
c-1-825-2-**40**-1-003	39.5	950	2.23	1.71	0.0029	500	0.005	710	1670
c-1-825-1-**40**-1-001	40.7	940	1.55	2.45	0.0020	1650	0.004	1290	1620
c-1-825-1-**40**-1-002	40.2	930	1.53	2.44	0.0019	1360	0.004	1170	1490
c-1-825-1-**40**-1-003	40.4	950	1.54	2.49	0.0020	1110	0.004	1020	1550
c-1-825-4-**40**-2-001	80.4	1730	4.36	1.59	0.0070	1440	0.011	1160	2500
c-1-825-4-**40**-2-002	81.6	1930	4.56	1.70	0.0067	1290	0.010	1200	3150
c-1-825-4-**40**-2-003	79.8	1870	4.42	1.70	0.0071	1410	0.011	1130	3010
c-1-825-2-**60**-1-001	38.9	1370	2.74	2.01	0.0059	1680	0.010	1220	3360
c-1-825-2-**60**-2-002	38.4	1350	2.69	2.01	0.0060	1710	0.010	1200	3400
c-1-825-2-**60**-1-003	38.9	1360	2.69	2.03	0.0059	1720	0.010		

Table 7. Gas Yields from Bookcase Material Tests

Code	Specimen Mass (g)	Primary Air flow (L/min)	Φ	y_{CO2} (g/g)	y_{CO} (g/g)	y_{HCN} (g/g)	y_{HCl} (g/g)	y_{NO} (g/g)	y_{NO2} (g/g)	$y_{acrolein}$ (g/g)	y_{form} (g/g)
b-1-650-10-40-1-001	14.8	11.6	0.43	1.38	< 0.0010	< 0.0018	< 0.0014	< 0.0039	< 0.0034	< 0.0021	< 0.0022
b-1-650-10-40-1-002	14.3	11.6	0.42	1.45	< 0.0011	< 0.0019	< 0.0015	< 0.0042	< 0.0037	< 0.0023	< 0.0024
b-1-650-10-40-1-003	14.8	11.6	0.44	1.46	< 0.0011	< 0.0018	< 0.0014	< 0.0041	< 0.0036	< 0.0022	< 0.0023
b-2-650-10-40-1-001	14.7	11.6	0.43	1.40	< 0.0011	< 0.0018	< 0.0014	< 0.0039	< 0.0035	< 0.0021	< 0.0023
b-2-650-10-40-1-002	13.7	11.6	0.41	1.53	< 0.0012	< 0.0020	< 0.0016	< 0.0045	< 0.0039	< 0.0024	< 0.0026
b-2-650-10-40-1-003	13.9	11.6	0.35	1.53	< 0.0012	< 0.0020	< 0.0016	< 0.0045	< 0.0039	< 0.0024	< 0.0026
b-1-650-13-40-1-001a	14.6	15.1	0.34	1.64	< 0.0011	< 0.0019	< 0.0015	< 0.0043	< 0.0037	< 0.0023	< 0.0024
b-1-650-13-40-1-002	14.2	15.1	0.35	1.68	< 0.0012	< 0.0020	< 0.0015	< 0.0044	< 0.0039	< 0.0024	< 0.0025
b-1-650-13-40-1-003	14.9	15.1	0.38	1.65	< 0.0011	< 0.0018	< 0.0014	< 0.0041	< 0.0036	< 0.0022	< 0.0023
b-1-650-20-40-2-001	28.1	22.2	0.48	1.69	< 0.0006	< 0.0010	< 0.0008	< 0.0022	< 0.0020	< 0.0012	< 0.0013
b-1-650-20-40-2-002	29.5	23.1	0.46	1.60	< 0.0006	< 0.0009	< 0.0007	< 0.0021	< 0.0018	< 0.0011	< 0.0012
b-1-650-20-40-2-003	31.4	23.1	0.44	1.46	< 0.0005	< 0.0009	< 0.0007	< 0.0020	< 0.0018	< 0.0011	< 0.0011
b-1-650-15-60-1-001	16.1	17.3	0.37	0.88	< 0.0007	< 0.0012	< 0.0010	< 0.0028	< 0.0024	< 0.0015	< 0.0016
b-1-650-15-60-1-002	14.3	17.4	0.41	0.73	< 0.0008	< 0.0014	< 0.0011	< 0.0030	< 0.0027	< 0.0016	< 0.0017
b-1-650-15-60-1-003	15.8	17.4	0.49	0.90	< 0.0008	< 0.0013	< 0.0010	< 0.0029	< 0.0025	< 0.0015	< 0.0016
b-1-825-2-40-1-001	15.5	2.61	1.89	0.51	0.095	< 0.0019	< 0.0015	< 0.0042	< 0.0037	< 0.0023	< 0.0024
b-1-825-2-40-1-002	14.8	2.57	1.89	0.28	0.048	< 0.0019	< 0.0015	< 0.0043	< 0.0038	< 0.0023	< 0.0025
b-1-825-2-40-1-003	15.8	2.57	1.93	0.28	0.044	< 0.0019	< 0.0015	< 0.0042	< 0.0037	< 0.0022	< 0.0024
b-2-825-2-40-1-001	15.1	2.57	1.87	0.29	0.032	< 0.0020	< 0.0015	< 0.0043	< 0.0038	< 0.0023	< 0.0025
b-2-825-2-40-1-002	13.5	2.57	1.72	0.30	0.039	< 0.0021	< 0.0016	< 0.0047	< 0.0041	< 0.0025	< 0.0027
b-2-825-2-40-1-003	15.5	2.57	1.94	0.29	0.049	< 0.0019	< 0.0015	< 0.0042	< 0.0037	< 0.0022	< 0.0024
b-1-825-1-40-1-001	15.6	1.76	2.71	0.24	0.074	< 0.0020	< 0.0015	< 0.0043	< 0.0038	< 0.0023	< 0.0025
b-1-825-1-40-1-002	14.8	1.77	2.67	0.21	0.049	< 0.0020	< 0.0015	< 0.0044	< 0.0039	< 0.0024	< 0.0025
b-1-825-1-40-1-003	15.1	1.77	2.77	0.20	0.067	< 0.0019	< 0.0015	< 0.0042	< 0.0037	< 0.0022	< 0.0024
b-1-825-4-40-2-001	26.9	5.00	1.75	0.34	0.050	< 0.0011	< 0.0008	< 0.0024	< 0.0021	< 0.0013	< 0.0014
b-1-825-4-40-2-002	28.8	5.00	1.83	0.32	0.067	< 0.0010	< 0.0008	< 0.0023	< 0.0020	< 0.0012	< 0.0013
b-1-825-4-40-2-003	29.1	5.00	1.85	0.28	0.053	< 0.0010	< 0.0008	< 0.0022	< 0.0020	< 0.0012	< 0.0013
b-1-825-3-60-1-001	15.7	3.53	2.12	0.25	0.052	< 0.0013	< 0.0010	< 0.0028	< 0.0025	< 0.0015	< 0.0016
b-1-825-3-60-1-002	13.8	3.39	1.97	0.29	0.052	< 0.0014	< 0.0011	< 0.0032	< 0.0028	< 0.0017	< 0.0018
b-1-825-3-60-1-003	14.3	3.38	1.97	0.29	0.044	< 0.0014	< 0.0011	< 0.0032	< 0.0028	< 0.0017	< 0.0018

Table 8. Gas Yields from Sofa Material Tests

Code	Specimen Mass (g)	Primary Airflow (L/min)	Φ	y_{CO2} (g/g)	y_{CO} (g/g)	y_{HCN} (g/g)	y_{HCl} (g/g)	y_{NO} (g/g)	y_{NO2} (g/g)	$y_{acrolein}$ (g/g)	y_{form} (g/g)
s-1-650-10-40-1-001	10.3	11.6	0.44	1.91	0.024	< 0.0027	< 0.0021	< 0.0060	< 0.0053	< 0.0032	< 0.0034
s-1-650-10-40-1-002	10.5	11.6	0.36	1.69	0.026	< 0.0026	< 0.0020	< 0.0057	< 0.0050	< 0.0030	< 0.0033
s-1-650-10-40-1-003	10.2	11.6	0.46	2.02	0.028	< 0.0028	< 0.0022	< 0.0062	< 0.0055	< 0.0033	< 0.0036
s-1-650-13-40-1-001	10.5	15.0	0.18	0.92	0.015	< 0.0027	< 0.0020	< 0.0059	< 0.0052	< 0.0032	< 0.0034
s-1-650-13-40-1-002	10.8	15.1	0.18	0.87	0.022	0.0029	< 0.0019	< 0.0056	< 0.0049	< 0.0030	< 0.0032
s-1-560-13-40-1-003	10.7	15.1	0.20	0.95	0.033	0.0039	< 0.0019	< 0.0056	< 0.0049	< 0.0030	< 0.0032
s-1-650-5-40-5-001	7.8	6.0	0.21	0.68	0.012	< 0.0034	< 0.0026	< 0.0076	< 0.0067	< 0.0041	< 0.0044
s-1-650-5-40-5-003	7.8	5.8	0.26	0.64	0.008	< 0.0035	< 0.0027	< 0.0077	< 0.0067	< 0.0041	< 0.0044
s-1-650-15-60-1-001	10.4	17.3	0.22	0.90	0.006	< 0.0018	< 0.0014	< 0.0041	< 0.0036	< 0.0022	< 0.0023
s-1-650-15-60-1-002	10.4	17.4	0.19	0.78	0.002	< 0.0018	< 0.0014	< 0.0040	< 0.0035	< 0.0021	< 0.0023
s-1-650-15-60-1-003	10.7	17.4	0.17	0.68	< 0.0010	< 0.0018	< 0.0014	< 0.0039	< 0.0034	< 0.0021	< 0.0022
s-1-825-2-40-1-001	10.5	2.92	1.25	0.54	0.142	0.0067	< 0.0020	< 0.0059	< 0.0052	< 0.0031	< 0.0034
s-1-825-2-40-1-002	13.7	2.92	1.08	0.59	0.144	0.0103	< 0.0019	< 0.0056	< 0.0049	< 0.0030	< 0.0032
s-1-825-2-40-1-003	9.7	2.92	1.89	0.31	0.173	0.0074	< 0.0024	< 0.0068	< 0.0060	< 0.0036	< 0.0039
s-1-825-1-40-1-001	10.4	2.01	1.88	0.35	0.161	0.0095	< 0.0019	< 0.0056	< 0.0049	< 0.0030	< 0.0032
s-1-825-1-40-1-002	10.4	2.00	1.64	0.38	0.165	0.0088	< 0.0020	< 0.0056	< 0.0049	< 0.0030	< 0.0032
s-1-825-1-40-1-003	10.2	2.18	2.14	0.16	0.075	0.0080	< 0.0021	< 0.006	< 0.0053	< 0.0032	< 0.0034
s-1-825-1-40-5-001a	7.4	1.18	1.83	0.20	0.078	< 0.0037	< 0.0029	< 0.0083	< 0.0073	< 0.0044	< 0.0048
s-1-825-1-40-5-002a	7.0	1.18	1.29	0.21	0.063	0.0061	< 0.0034	< 0.0097	< 0.0085	< 0.0052	< 0.0056
s-1-825-1-40-5-003a	6.1	1.65	2.67	0.26	0.147	0.0054	< 0.0034	< 0.0099	< 0.0087	< 0.0053	< 0.0057
s-1-825-2-60-1-001	13.5	1.88	2.84	0.30	0.142	0.0099	< 0.0015	< 0.0042	< 0.0037	< 0.0023	< 0.0024
s-1-825-2-60-1-002b	11.0	1.89	2.76	0.30	0.159	0.0096	< 0.0014	< 0.004	< 0.0035	< 0.0021	< 0.0023
s-1-825-2-60-1-003	10.4	1.88				No FTIR data					

Table 9. Gas Yields from Cable Material Tests

Code	Specimen Mass (g)	Primary Airflow (L/min)	Φ	y_{CO2} (g/g)	y_{CO} (g/g)	y_{HCN} (g/g)	y_{HCl} (g/g)	y_{NO} (g/g)	y_{NO2} (g/g)	$y_{acrolein}$ (g/g)	y_{form} (g/g)
c-1-650-10-40-1-001	39.2	11.6	0.36	1.23	0.071	<0.0024	0.26	<0.0054	<0.0048	<0.0029	<0.0031
c-1-650-10-40-1-002	39.3	11.6	0.36	1.84	0.098	<0.0037	0.38	<0.0081	<0.0071	<0.0043	<0.0047
c-1-650-10-40-1-003	39.1	11.6	0.34	1.13	0.095	<0.0024	0.25	<0.0054	<0.0048	<0.0029	<0.0031
c-2-650-10-40-1-001	39.2	11.5	0.40	1.36	0.083	<0.0024	0.31	<0.0054	<0.0047	<0.0029	<0.0031
c-2-650-10-40-1-002	39.6	11.6	0.38	1.37	0.099	<0.0026	0.31	<0.0058	<0.0051	<0.0031	<0.0033
c-2-650-10-40-1-003a	39.4	11.6	0.37	1.24	0.090	<0.0024	0.27	<0.0054	<0.0047	<0.0029	<0.0031
c-1-650-13-40-1-001	39.1	15.1	0.26	1.14	0.065	<0.0025	0.24	<0.0055	<0.0048	<0.0029	<0.0031
c-1-650-13-40-1-002	38.7	14.3	0.29	1.24	0.078	<0.0026	0.25	<0.0057	<0.005	<0.0031	<0.0033
c-1-650-13-40-1-003	39.2	15.1	0.26	1.14	0.068	<0.0025	0.26	<0.0055	<0.0049	<0.0030	<0.0032
c-1-650-20-40-2-001	77.7	22.8	0.32	1.21	0.074	<0.0013	0.18	<0.0029	<0.0026	<0.0016	<0.0017
c-1-650-20-40-2-002	79	23.1	0.32	1.17	0.063	<0.0012	0.18	<0.0027	<0.0024	<0.0015	<0.0016
c-1-650-20-40-2-003	77.3	23.1	0.30	1.08	0.075	<0.0012	0.21	<0.0027	<0.0024	<0.0015	<0.0016
c-1-650-15-60-1-001	38.6	17.4	0.32	1.18	0.066	<0.0018	0.24	<0.0039	<0.0034	<0.0021	<0.0022
c-1-650-15-60-1-002	38.8	17.4	0.33	1.23	0.066	<0.0017	0.25	<0.0039	<0.0034	<0.0021	<0.0022
c-1-650-15-60-1-003	38.6	17.4	0.32	1.19	0.064	<0.0017	0.23	<0.0039	<0.0034	<0.0021	<0.0022
c-1-650-10-20-2-001	76.4	10.0	0.50	1.52	0.094	<0.0023	0.29	<0.0050	<0.0044	<0.0027	<0.0029
c-1-650-10-20-2-002	75.4	10.0	0.47	1.43	0.091	<0.0023	0.30	<0.0051	<0.0045	<0.0027	<0.0029
c-1-650-10-20-2-003	76.5	10.0	0.48	1.46	0.090	<0.0023	0.30	<0.0050	<0.0044	<0.0027	<0.0029
c-1-825-2-40-1-001	39.6	2.23	1.71	0.29	0.036	<0.0024	0.16	<0.0054	<0.0047	<0.0029	<0.0031
c-1-825-2-40-1-002	39.8	2.23	1.70	0.37	0.049	<0.0025	0.14	<0.0055	<0.0048	<0.0029	<0.0031
c-1-825-2-40-1-003	40	2.23	1.71	0.34	0.044	<0.0024	0.11	<0.0054	<0.0047	<0.0029	<0.0031
c-1-825-2-40-1-001	39.1	2.26	1.68	0.33	0.041	<0.0024	0.13	<0.0054	<0.0048	<0.0029	<0.0031
c-1-825-2-40-1-002	38.6	2.23	1.69	0.27	0.034	<0.0024	0.14	<0.0054	<0.0047	<0.0029	<0.0031
c-1-825-2-40-1-003	39.5	2.23	1.71	0.32	0.036	<0.0024	0.15	<0.0053	<0.0047	<0.0028	<0.0031
c-1-825-1-40-1-001	40.7	1.55	2.45	0.22	0.118	<0.0024	0.15	<0.0053	<0.0047	<0.0028	<0.0030
c-1-825-1-40-1-002	40.2	1.53	2.44	0.22	0.098	<0.0024	0.14	<0.0054	<0.0047	<0.0029	<0.0031
c-1-825-1-40-1-003	40.4	1.54	2.49	0.22	0.078	<0.0024	0.14	<0.0053	<0.0046	<0.0028	<0.0030
c-1-825-4-40-2-001	80.4	4.36	1.59	0.44	0.058	<0.0014	0.13	<0.0030	<0.0026	<0.0016	<0.0017
c-1-825-4-40-2-002	81.6	4.56	1.70	0.38	0.046	<0.0012	0.15	<0.0027	<0.0023	<0.0014	<0.0015
c-1-825-4-40-2-003	79.8	4.42	1.70	0.41	0.052	<0.0012	0.14	<0.0027	<0.0024	<0.0015	<0.0016
c-1-825-2-60-1-001	38.9	2.74	2.01	0.48	0.088	<0.0018	0.23	<0.0039	<0.0034	<0.0021	<0.0022
c-1-825-2-60-2-002	38.4	2.69	2.01	0.50	0.090	<0.0018	0.23	<0.0040	<0.0035	<0.0021	<0.0023
c-1-825-2-60-1-003	38.9	2.69	2.03	0.48	0.091	No FTIR data					

25

B. CALCULATIONS OF TOXIC GAS YIELDS WITH UNCERTAINTIES

Table 10 contains the yields of the combustion products calculated using the data from Table 7 Table 8, and Table 9. The estimated uncertainties reflect the repeatability of the volume fractions in replicate tests, uncertainties in the other terms in the yields calculations, and degree of proximity of the measured values to the background levels. The Fire Stage column refers to pre-flashover and post-flashover, which is a better description of the conditions in the full scale experiments. For this work we are considering the well-ventilated mode of the tube furnace to be equivalent to pre-flashover.

Table 10. Yields of Combustion Products from Tube Furnace Tests.

Gas	Fire Stage		Bookcase			Sofa			Cable		
CO_2	Pre	Solid	1.43	±	3.2 %	1.87	±	8.9 %	1.40	±	27 %
		Diced	1.49	±	5.0 %		X		1.33	±	5.5 %
		Air x1.3	1.66	±	1.2 %	0.91	±	4.6 %	1.17	±	4.8 %
		Mass x2	1.58	±	7.2 %	0.66	±	4.5 %	1.15	±	5.5 %
		Feed x1.5	0.83	±	11 %	0.79	±	14 %	1.20	±	2.1 %
	Post	Solid	0.36	±	38 %	0.56	±	6.6 %	0.33	±	12 %
		Diced	0.29	±	1.9 %		X		0.31	±	11 %
		Air x0.7	0.21	±	8.8 %	0.35	±	9.6 %	0.22	±	0.1 %
		Mass x2	0.31	±	10 %	0.23	±	18 %	0.41	±	8.2 %
		Feed x1.5	0.28	±	8.2 %	0.28	±	7.1 %	0.49	±	1.7 %
CO	Pre	Solid	<		0.0011	0.026	±	10.1 %	0.088	±	16 %
		Diced	<		0.0012		X		0.091	±	8.8 %
		Air x1.3	<		0.0011	0.023	±	38.1 %	0.070	±	10 %
		Mass x2	<		0.0006	0.010	±	31.0 %	0.071	±	9.3 %
		Feed x1.5	<		0.0008	0.003	±	88.2 %	0.065	±	1.6 %
	Post	Solid	0.062	±	45 %	0.143	±	1.4 %	0.043	±	14 %
		Diced	0.040	±	21 %		X		0.037	±	9.5 %
		Air x0.7	0.063	±	21 %	0.166	±	3.6 %	0.098	±	20 %
		Mass x2	0.057	±	16 %	0.052	±	65.1 %	0.052	±	11 %
		Feed x1.5	0.050	±	8.9 %	0.143	±	10.4 %	0.090	±	1.7 %
HCN	Pre	Solid	<		0.0018	<		0.0027	<		0.0028
		Diced	<		0.0019		X		<		0.0025
		Air x1.3	<		0.0019	0.0032	±	30 %	<		0.0025
		Mass x2	<		0.0009	<		0.0035	<		0.0012
		Feed x1.5	<		0.0013	<		0.0018	<		0.0017
	Post	Solid	<		0.0019	0.0081	±	33 %	<		0.0024
		Diced	<		0.0020		X		<		0.0024
		Air x0.7	<		0.0020	0.0088	±	19 %	<		0.0024
		Mass x2	<		0.0010	0.0058	±	19 %	<		0.0013
		Feed x1.5	<		0.0014	0.0098	±	12 %	<		0.0018

Gas	Fire Stage		Bookcase		Sofa		Cable		
HCl	Pre	Solid	<	0.0014	<	0.0021	0.30	±	34 %
		Diced	<	0.0015	X		0.30	±	18 %
		Air x1.3	<	0.0015	<	0.0019	0.25	±	14 %
		Mass x2	<	0.0007	<	0.0027	0.19	±	19 %
		Feed x1.5	<	0.0010	<	0.0014	0.24	±	14 %
	Post	Solid	<	0.0015	<	0.0021	0.14	±	29 %
		Diced	<	0.0015	X		0.14	±	19 %
		Air x0.7	<	0.0015	<	0.0020	0.14	±	14 %
		Mass x2	<	0.0008	<	0.0033	0.14	±	16 %
		Feed x1.5	<	0.0011	<	0.0015	0.23	±	12 %
NO	Pre	Solid	<	0.0041	<	0.0060	<	0.0063	
		Diced	<	0.0043	X		<	0.0055	
		Air x1.3	<	0.0043	<	0.0057	<	0.0056	
		Mass x2	<	0.0021	<	0.0077	<	0.0028	
		Feed x1.5	<	0.0029	<	0.0040	<	0.0039	
	Post	Solid	<	0.0042	<	0.0061	<	0.0054	
		Diced	<	0.0044	X		<	0.0054	
		Air x0.7	<	0.0043	<	0.0057	<	0.0053	
		Mass x2	<	0.0023	<	0.0096	<	0.0028	
		Feed x1.5	<	0.0031	<	0.0041	<	0.0040	
NO_2	Pre	Solid	<	0.0036	<	0.0053	<	0.0056	
		Diced	<	0.0038	X		<	0.0048	
		Air x1.3	<	0.0037	<	0.0050	<	0.0049	
		Mass x2	<	0.0019	<	0.0067	<	0.0025	
		Feed x1.5	<	0.0025	<	0.0035	<	0.0034	
	Post	Solid	<	0.0037	<	0.0054	<	0.0047	
		Diced	<	0.0039	X		<	0.0047	
		Air x0.7	<	0.0038	<	0.0050	<	0.0047	
		Mass x2	<	0.0020	<	0.0084	<	0.0024	
		Feed x1.5	<	0.0027	<	0.0036	<	0.0035	
Acrolein	Pre	Solid	<	0.0022	<	0.0032	<	0.0034	
		Diced	<	0.0023	X		<	0.0030	
		Air x1.3	<	0.0023	<	0.0031	<	0.0030	
		Mass x2	<	0.0011	<	0.0041	<	0.0015	
		Feed x1.5	<	0.0015	<	0.0021	<	0.0021	
	Post	Solid	<	0.0023	<	0.0032	<	0.0029	
		Diced	<	0.0023	X		<	0.0029	
		Air x0.7	<	0.0023	<	0.0031	<	0.0028	
		Mass x2	<	0.0012	<	0.0052	<	0.0015	
		Feed x1.5	<	0.0016	<	0.0022	<	0.0021	

Gas	Fire Stage		Bookcase	Sofa	Cable
Formaldehyde	Pre	Solid	< 0.0023	< 0.0034	< 0.0036
		Diced	< 0.0025	X	< 0.0032
		Air x1.3	< 0.0024	< 0.0033	< 0.0032
		Mass x2	< 0.0012	< 0.0044	< 0.0016
		Feed x1.5	< 0.0016	< 0.0023	< 0.0022
	Post	Solid	< 0.0024	< 0.0035	< 0.0031
		Diced	< 0.0025	X	< 0.0031
		Air x0.7	< 0.0025	< 0.0033	< 0.0030
		Mass x2	< 0.0013	< 0.0055	< 0.0016
		Feed x1.5	< 0.0017	< 0.0024	< 0.0023

The values in Table 11 are the values from Table 10 divided by the notional yields from Table 3. Thus the uncertainties are the combined uncertainties from those two tables.

Table 11. Fractions of Notional Yields.

Gas	Fire Stage		Bookcase			Sofa			Cable		
CO_2	Pre	Solid	0.83	±	5 %	0.96	±	13 %	0.66	±	28 %
		Diced	0.86	±	6 %		X		0.63	±	7 %
		Air x1.3	0.96	±	3 %	0.47	±	9 %	0.56	±	6 %
		Mass x2	0.92	±	9 %	0.34	±	9 %	0.55	±	7 %
		Feed x1.5	0.49	±	12 %	0.40	±	18 %	0.57	±	4 %
	Post	Solid	0.21	±	39 %	0.29	±	11 %	0.16	±	13 %
		Diced	0.17	±	3 %		X		0.15	±	12 %
		Air x0.7	0.12	±	10 %	0.18	±	14 %	0.10	±	2 %
		Mass x2	0.18	±	11 %	0.12	±	22 %	0.19	±	10 %
		Feed x1.5	0.16	±	10 %	0.14	±	11 %	0.23	±	3 %
CO	Pre	Solid	< 0.0010			0.021	±	14 %	0.066	±	17 %
		Diced	< 0.0011				X		0.068	±	10 %
		Air x1.3	< 0.0010			0.019	±	42 %	0.053	±	11 %
		Mass x2	< 0.0005			0.008	±	35 %	0.053	±	10 %
		Feed x1.5	< 0.0007			0.002	±	92 %	0.049	±	3 %
	Post	Solid	0.057	±	46 %	0.115	±	6 %	0.032	±	15 %
		Diced	0.037	±	22 %		X		0.028	±	11 %
		Air x0.7	0.058	±	22 %	0.134	±	8 %	0.074	±	21 %
		Mass x2	0.052	±	17 %	0.042	±	69 %	0.039	±	12 %
		Feed x1.5	0.045	±	10 %	0.115	±	14 %	0.067	±	3 %

Gas	Fire Stage		Bookcase			Sofa			Cable		
HCN	Pre	Solid	<	0.032			<	0.014		<	0.071
		Diced	<	0.034			X			<	0.062
		Air x1.3	<	0.033		0.0164	±	34 %		<	0.063
		Mass x2	<	0.016			<	0.018		<	0.031
		Feed x1.5	<	0.023			<	0.009		<	0.043
	Post	Solid	<	0.033		0.0421	±	37 %		<	0.061
		Diced	<	0.035			X			<	0.060
		Air x0.7	<	0.035		0.0454	±	23 %		<	0.060
		Mass x2	<	0.018		0.0298	±	23 %		<	0.032
		Feed x1.5	<	0.024		0.0505	±	17 %		<	0.045
HCl	Pre	Solid	<	0.55			<	0.30	0.90	±	36 %
		Diced	<	0.59			X		0.90	±	18 %
		Air x1.3	<	0.56			<	0.28	0.75	±	14 %
		Mass x2	<	0.28			<	0.38	0.58	±	20 %
		Feed x1.5	<	0.40			<	0.20	0.73	±	13 %
	Post	Solid	<	0.58			<	0.30	0.41	±	29 %
		Diced	<	0.59			X		0.43	±	19 %
		Air x0.7	<	0.58			<	0.29	0.43	±	14 %
		Mass x2	<	0.31			<	0.48	0.42	±	16 %
		Feed x1.5	<	0.41			<	0.21	0.70	±	12 %

This page intentionally left blank

V. DISCUSSION

A. OVERALL TEST VALUES

The principal outcome of this series of tests is a well-documented set of combustion product yields. This includes the numerical values themselves, the apparatus conditions under which they were obtained, the uncertainty in their calculated values, and the repeatability of the tests.

Next most important is a determination of the extent to which the toxic gas yields are affected by variations in the test protocol that are reasonable in light of possible variations in combustion conditions in fires involving the intact products.

Third, it is important to evaluate the quality of the derived knowledge in the context of its intended use. The yield information would be used with a computational fire model (zone or CFD) to generate the time-dependent environment generated by a fire. Equations such as those in ISO 13571[9] would then be used to assess whether the combination of occupancy design, contained combustibles, and occupant/responder characteristics lead to the desired level of life safety.

The documentation of the yields has been provided in the earlier sections. The following examines the context and quality of the results.

B. SPECIMEN PERFORMANCE AND TEST REPEATABILITY

1. General observations

The bookcase and cable specimens were relatively easy to handle, but the sofa material proved to be more challenging. First, the low density of the foam made diced pieces prone to blowing around to the degree that we were never successful in conducting a test of the diced material without some fraction of the specimen blowing out of the sample boat and into the hot part of the furnace. Therefore, we do not report any results for diced sofa material. Second, the low density resulted in a relatively large specimen, one that just fit the dimensions of the furnace. Therefore, it was not possible to double the mass loading; instead we examined the effect of halving the mass loading. The third challenge with the foam was that once ignited, its high flammability resulted in a flame spread rate that exceeded the feed rate into the furnace. Thus, the flaming and initial decomposition took place upstream of the furnace (Figure 8). When ignited, polyurethane foams are known to undergo decomposition into TDI, which burns off initially, and a polyol residue, which requires more thermal input before it ignites.[31] Therefore, during the majority of the tests, there were actually two flame zones, one upstream of the furnace resulting from the combustion of the TDI, and one well within the furnace resulting from the combustion of the polyol. The products from the first flame pass through the second one, possibly undergoing further reaction, and then the products of both flames travelled together into the dilution and sampling chamber.

Figure 8. Close-up of the sofa specimen burning upstream of the furnace

A similar two-stage reaction can be seen for the bookcase specimens. A moderate amount of residue is left behind after flaming combustion, and this material continues to be exposed to high temperatures downstream from the flame, potentially producing additional products in proportion to how much residue is present in the post-flame region. This contribution may rise as the experiment progresses due to the increasing length of the residue in the downstream section of the furnace, and due to the chemistry of the residue differing from that of the virgin specimen.

2. CO_2 and CO

Figure 9 shows a typical record for a single experiment. Several important features can be observed. The relatively larger spike in CO at the start of the test is the result of an instantaneous or impulse release of CO as the specimen first ignites, broadened by the characteristic mixing time of the dilution chamber. (If modeled as a perfectly-stirred reactor, the chamber has a time constant of 15 s, resulting in a decay time of a little over 2 minutes to reach 1 % of the initial value.) Because the intent of this apparatus is to determine the "steady state" generation of toxic gases, this initial time period is therefore discarded from the analysis. A similar peak in gas concentration can often be found at the end of the experiment, which we attribute to a) the lack of additional unburnt material to conduct heat away from the fire, and b) the end face of the specimen becoming involved in the fire, increasing the surface area and fire size. Therefore the period of interest for determining the "steady state" production lies between these two peaks. (Note that the mass loss rate may or may not be higher during ignition and extinction; if it is then the yields are somewhat underestimated.)

However, as can be seen in Figure 9, the gas concentrations continue to vary between the ignition and extinction times, more so than allowed by the ISO/TS. We examined the influence of the data window selected for averaging on the reported value, and fond it to be considerably smaller than the repeatability-related uncertainty. Because of the mixing chamber broadening of instantaneous events, several hundred seconds of data were used.

As intended, changing the temperature and equivalence ratio caused a dramatic decrease in CO_2 yields. In all comparable cases the yield decreased by a factor of 3 to 4, for example with the bookcase material from 1.4 to 0.3. At the same time, this caused a substantial increase in CO yields for the bookcase and sofa materials, but generally a decrease in CO yields for the cable material (about a factor of 2). For the bookcase material, the CO was below the detection limit under well ventilated conditions, but around 0.05 for post-flashover conditions, increasing by at least a factor of 50. For the sofa material, the CO yield increases by a factor of 5 under postflashover conditions, to about 0.14.

The other variations we examined produced relatively minor changes to the CO and CO_2 yields when compared to their corresponding well ventilated and postflashover conditions. The biggest changes in CO_2 yields come when the sofa material is burned under well-ventilated conditions at half the mass loading, causing the yield to fall by 65 %. When the cable material was fed at the higher feed rate, its CO_2 yield increased by 45 %.

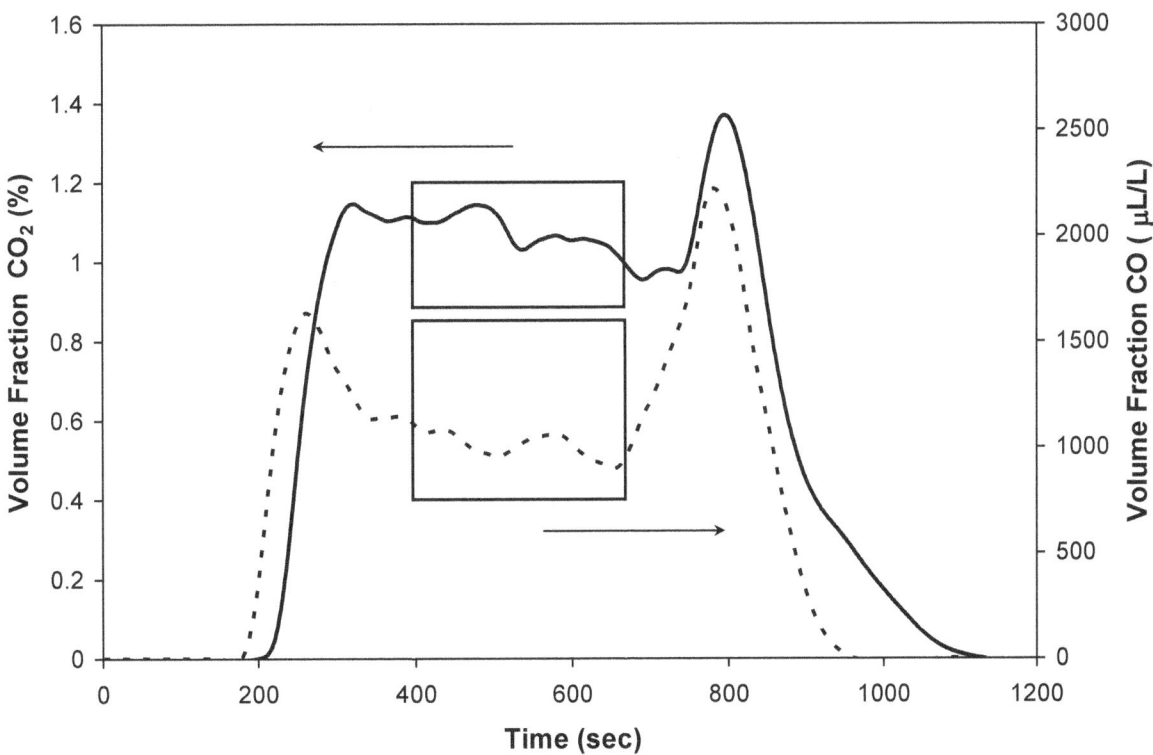

Figure 9. Measured CO (dotted) and CO_2 (solid) Volume Fractions in the Dilution Chamber When Burning a PVC-clad Electrical Cable. Rectangles indicate the time period from which the average volume fraction is determined.

The biggest changes in CO yield were for the sofa material burned under well ventilated conditions, where increasing the feed rate decreased the CO yield by 90 %, and for the cable material under postflashover conditions, where a further increase in the equivalence ratio (decrease in air) resulted in the CO yield more than doubling. This effect was much lower for the other two specimen types. In looking at the effects of variations on the standard conditions no strong trends emerge—the degree and direction is specific to specimen type and fire stage.

In terms of repeatability, most scenarios resulted in CO_2 yields that were repeatable to under a 10 % standard deviation. For those that had a higher variability, the cause was usually one of the three runs having substantially different gas yields, all other things being essentially equal. For example, for the bookcase material at standard post-flashover conditions (coded b-1-825-2-40-1 in Table 4 and Table 7, the first run has considerably higher gas concentrations than the second and third, even though masses, flow rates, and equivalence ratios are all very close. Figure 10 shows the raw data from these three experiments, and it is clear that one of them simply produced more CO_2, (and CO), even though the mass loss and equivalence ratio were nearly identical to those in the other tests.

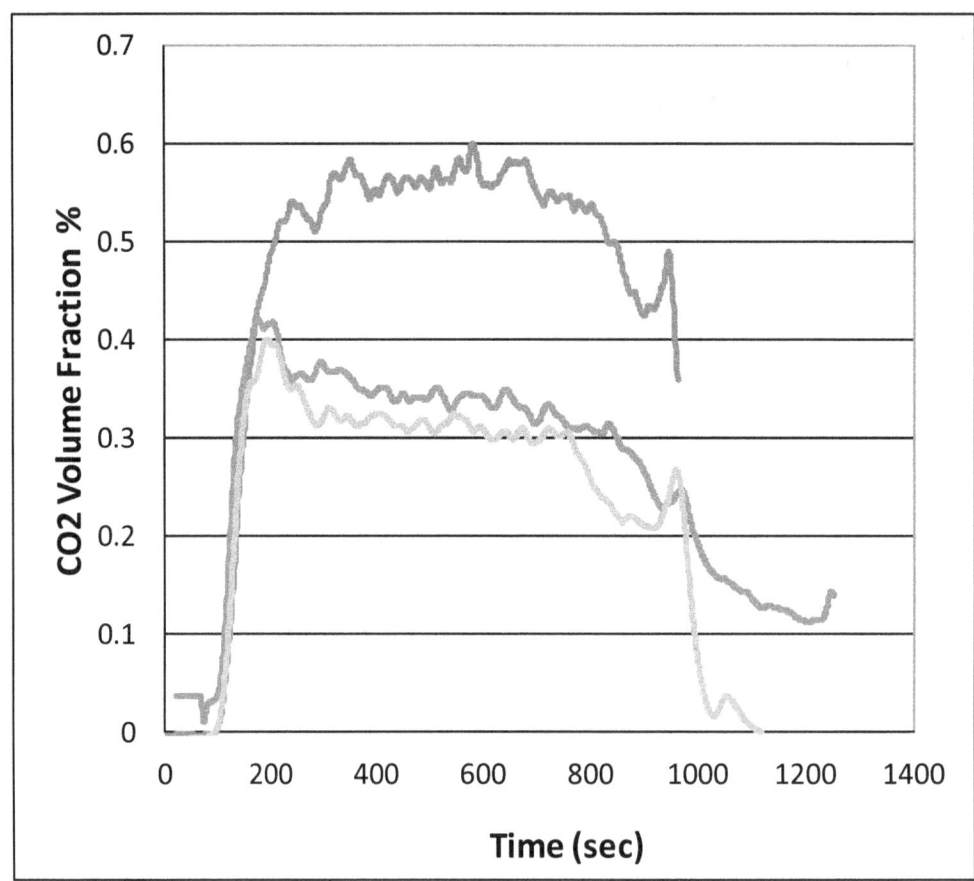

Figure 10. Raw data from Bookcase Material Test b-1-825-2-40.

The other condition with high variability in CO_2 yield was the cable material burned under standard well ventilated combustion. Again, the data in Table 6 reveals that, all other things being equal, the measured mass loss rate for the second experiment was considerably lower than in the other two, for the same gas concentrations, resulting in higher yields.

The variability of CO yields was somewhat higher, averaging around 20 % for the bookcase, 20 % to 40 % for the sofa (post- and pre-flashover respectively), and 10 % for the cable. If the highest variable yields from the sofa are omitted, then the average variability falls to 5 % and

26 % for postflashover and well ventilated respectively. The high variability in the sofa values comes mainly from two cases: well ventilated with the feed rate increased 50 %, and postflashover with the mass load halved. In the first case, the cause is very low CO yields, close to the limit of detection (below, in fact, for one of the triplicates). This is a case of "small differences between small numbers have large relative importance." In the postflashover case, looking back to Table 5 and Table 8, we see that at the time of the experiments, the results were sufficiently variable that we conducted additional runs.

3. HCl and HCN

HCN was detected via FTIR and was only observed in the sofa tests. Under well ventilated conditions, it was only observed when the primary air flow was increased by 30 %. It was observed in only 2 of the 3 replicate experiments, and at concentrations only slightly over the minimum detection limit. HCN was consistently observed under postflashover conditions, resulting in yields that were 2 to 5 times greater than the minimum detection limit and nearly 3 times the single yield measured from well ventilated combustion. The only other variation to have a significant effect was reducing the mass loading: as with the CO_2 and CO, halving the mass of foam resulted in a considerable decrease in yield (30 % here).

HCl was detected via FTIR and was only observed in the cable tests. Under well ventilated conditions, it was observed at yields between 0.2 and 0.3, which are close to the notional yields. The highest yield of HCl is obtained when the specimen is diced, while the lowest occurs when the fuel load is doubled. Under postflashover conditions, the yield is reduced by 40 %. The only operational variation that has an effect is to increase the feed rate by 50 %, which restores the yield to the well-ventilated value.

C. MEASURED VS. NOTIONAL VALUES

During sustained and complete combustion, the yield of CO_2 should approach its notional values, since CO_2 is a marker for combusted carbon. Under well ventilated conditions, the sofa material most closely approaches complete combustion with a yield of CO_2 that is 96 % of the notional yield. Under the same conditions, the yield of CO_2 from the bookcase material reaches 83 % of the notional yield, but this rises to 96 % when the flow of air is increased by 30 %. By this measure, the cable material experiences the least complete combustion, its yield of CO_2 reaching a maximum of 66 % of the notional yield under well ventilated conditions. Under postflashover conditions, the CO_2 yields for all specimens and conditions range from 10 % to 29 % of the notional yields.

The yields of CO from all the specimens ranged from 0.003 to 0.166. These values are consistent with combustion ranging from fuel-lean to fuel-rich[32] and are in some cases approach the 0.2 value expected of postflashover fires.[14]

The yields of HCl from the cable specimens approach their notional values. The deficit may reflect scavenging by the calcium carbonate filler in the cable jacket or wall losses. The detection limited yields of HCl from the sofa specimens was around 30 % of the notional yield under all conditions. The detection limited yield of HCl from the bookcase was around half the notional yield.

Yields of HCN from the sofa are between 2 % and 5 % of their notional values. The highest relative yield is found at postflashover conditions with the feed rate increased by 50 %.

D. SPECIES SAMPLING AND MEASUREMENT

1. Species Measurement Using FTIR Spectroscopy

FTIR spectroscopic analysis of combustion products has become fairly common in fire research laboratories. However, that does not mean that its use is straightforward. The data from a recent round robin involving FTIR measurement of toxic combustion products from a standardized apparatus showed interlaboratory variations of up to an order of magnitude. There are documents under development in ISO TC92 SC1 and SC3 to standardize the implementation.

We were able to obtain usable information using this technique. There are a number of lessons emerging from this test series that can provide useful input to these efforts, such as the following:

- The application of FTIR spectroscopy to fire testing requires the constant attention of an experienced professional at a level well beyond the demands of the more traditional fire test instrumentation.

- To maximize the opportunity for obtaining time resolved concentration data, we selected a small volume cell of short optical path length and operated without a soot filter. While some cleaning was necessary, it was not a major impediment. However, the short path length did limit the sensitivity, and moderated our ability to determine toxicologically important levels of the major gases.

- For future work, a longer path length should be considered. Ideally the test method achieves steady state in which case the response and residence time of the cell is unimportant. In reality, when real specimens are burned there are fluctuations on the order of seconds. This is ameliorated by the well-mixed nature of the dilution and sampling chamber, which has a characteristic time on the order of 60 sec.

- A heated sample line (as recommended in the SAFIR report[33] and ISO 19702[34]) enabled quantitative collection of HCl, a compound that is generally regarded as difficult to determine.

E. IMPORTANCE OF UNDETECTED GASES

The equations in ISO 13571 include provision for additional gases to be included in estimating the time available for escape or refuge from a fire: HBr, HF, SO_2, NO_2, acrolein (C_3H_4O) and formaldehyde (H_2CO). There was no Br, F, or S in any of the products examined in this project, so the first three of these gases were not expected. The presence of the latter three was not detected, thus establishing the upper limits of their presence at the volume fractions listed in Table 2.

To put the potential contributions of the sensory irritant gases (HCl, NO_2, acrolein, and formaldehyde) in context, we use the equations in ISO 13571 for calculating the Fractional Effective Dose (FED) for the narcotic gases, CO_2 and CO, and the Fractional Effective Concentration (FEC) for the four sensory irritant gases.

The FED equation is:

$$FED = \left[\sum_{t1}^{t2} \frac{\varphi_{CO}}{35\,000} \Delta t + \sum_{t1}^{t2} \frac{\varphi_{HCN}^{2.36}}{1.2 \cdot 10^6} \Delta t \right]^{\exp\left[\frac{\varphi_{CO_2}}{5}\right]},$$

where Δt is the exposure interval in minutes.

The FEC of the four irritant gases were estimated from their volume fractions and the incapacitating levels in ISO 13571 (F_{HCl}, etc.). The results are compiled in Table 12.

The FEC equation in ISO 13571 is:

$$FEC = \frac{\varphi_{HCl}}{F_{HCl}} + \frac{\varphi_{HBr}}{F_{HBr}} + \frac{\varphi_{HF}}{F_{HF}} + \frac{\varphi_{SO_2}}{F_{SO_2}} + \frac{\varphi_{NO_2}}{F_{NO_2}} + \frac{\varphi_{acrolein}}{F_{acrolein}} + \frac{\varphi_{formaldehyde}}{F_{formaldehyde}} + \sum \frac{\varphi_{irritant}}{F_{C_i}}$$

Table 12. Limits of Importance of Undetected Toxicants

	Volume fraction (μL/L)				Fractional FEC Contribution			
	HCl	NO$_2$	C$_3$H$_4$O	H$_2$CO	HCl	NO$_2$	C$_3$H$_4$O	H$_2$CO
Incapacitating Level	1000	250	30	250				
Bookcase	< 20	< 40	< 20	< 40	< 0.02	< 0.16	< 0.66	< 0.16
Sofa	< 20	< 40	< 20	< 40	< 0.02	< 0.16	< 0.66	< 0.16
Cable	4470	< 40	< 20	< 40	0.82	< 0.03	< 0.12	< 0.03
	to 1140	< 40	< 20	< 40	0.54	< 0.08	< 0.31	< 0.08

It stands out that the FEC contribution from acrolein is as much as two-thirds of an incapacitating level. This is because (a) the limit of detection is close to the listed incapacitating level and (b) the incapacitating level is very low. While there is agreement among experts that this value of 30 μL/L is reasonable, there are data that suggest strongly that this is unnecessarily conservative. Kaplan and co-workers exposed individual baboons to various concentrations of acrolein in air.[35] At the end of 5 min, each baboon was given a signal and could perform an action that led to escape from the test chamber. The baboons exposed at up to 500 μL/L escaped and survived. Those exposed to higher levels escaped, but died later. These data suggest that people should be able to accommodate a nearly instantaneous exposure to, e.g., at least 300 μL/L without becoming incapacitated. If we increase the incapacitating level of acrolein to 250 μL/L, the relative contribution to FEC become those in Table 13. In this case the relative importance of acrolein falls below 0.05 when other irritants are detected.

Table 13. Limits of Importance of Undetected Toxicants with Adjusted Acrolein Contribution

	Volume fraction (μL/L)				Fractional FEC Contribution			
	HCl	NO$_2$	C$_3$H$_4$O	H$_2$CO	HCl	NO$_2$	C$_3$H$_4$O	H$_2$CO
Incapacitating Level	1000	250	250	250				
Bookcase	< 20	< 40	< 20	< 40	< 0.05	< 0.38	< 0.19	< 0.38
Sofa	< 20	< 40	< 20	< 40	< 0.05	< 0.38	< 0.19	< 0.38
Cable	4470	< 40	< 20	< 40	0.92	< 0.03	< 0.02	< 0.03
	1140	< 40	< 20	< 40	0.74	< 0.10	< 0.05	< 0.10

Where specimens like the bookcase and sofa produce narcotic gases, CO and HCN, and irritants are below the limits of detection, the narcotic gases are the primary contributors to incapacitation. Assuming that the irritant gases are present in just under the limits of detection, in order for them to reach an incapacitating concentration, the relative concentration of CO would be incapacitating in a few minutes; the relative concentration of HCN would be incapacitating in under a minute.

VI. CONCLUSION

This paper reports toxic gas yield data for specimens cut from three complex combustibles: a bookcase, a sofa, and residential electrical power cable. The physical fire model used was the ISO/TS 19700 tube furnace. This apparatus is intended primarily for thermoplastics, but has some limited ability to test more complex assemblies. By design the equivalence ratio can be varied from fuel-lean to fuel-rich in order to simulate both well ventilated and postflashover fire stages. In addition to performing the tests under the standard conditions, this work examined the effect of varying the air flow rate, fuel feed rate, fuel load, and the effect of dicing the test specimens, to estimate the effects of the test variables and the conformation of the test specimen.

The findings were as follows:

- For the specimen types we investigated, the gas concentrations did not reach a steady state as defined in the technical specification. Therefore, we had to verify that our selection of time interval did not bias the result.

- The selection of data from the middle of the run, and the assumption of steady state may result in underestimation of the gas yields, as the burning rate appears higher at the beginning and end of the runs. In the absence of real time mass loss data, it might be more reliable to integrate the gas concentrations and convert to the mass of gas evolved.

- The combustion environment varies along the length of the furnace. At the peak temperature (925 K), the radiant flux is over 50 kW/m^2, with sharp fall-offs upstream and downstream. For specimens that burn in multiple stages (e.g. polyurethane), the less flammable component may proceed downstream of the initial reaction zone, and experience a higher equivalence ratio and temperature.

- Test specimens with high linear burning rates overtake the feed mechanism, therefore having a higher mass loss rate than intended. This is not addressed in the Technical Specification.

- The CO_2 yields were fairly repeatable and under well ventilated conditions represented much of the carbon in the specimens. The cable had the lowest fraction of carbon converted to CO_2 and a relatively higher yield of CO, indicating incomplete combustion even under well ventilated conditions. Under postflashover conditions, the CO_2 yields decreased considerably, with only 10 % to 30 % of the carbon in the fuel converted to CO_2.

- The CO yields are less repeatable, and there is no direct correlation between the repeatability of the CO_2 and CO yields. Between well ventilated and postflashover conditions, CO yields increased by a factor of 5 for the bookcase and sofa materials, but decreased by a factor of 2 for the cable materials. Only in the case of the sofa material under postflashover conditions do the CO yields approach the value of 0.2 found in real-scale post-flashover fires.

- The sum of the CO_2 and CO yields frequently accounted for half or less of the carbon originally in the specimen. Presumably, the post-test residue is enriched in carbon. As a result, the gaseous combustion products cannot be used to estimate the mass burning rate.

- The HCN yields were below the limits of detection for the bookcase and cable specimens. Even for the sofa materials, only one of the four well ventilated test conditions (increased air flow) resulted in sufficient HCN for detection. Under postflashover conditions HCN was detected at yields around 4 times the detection-limited yield. This represented 3 % to 5 % of the available nitrogen in the polyurethane foam.

- The HCl volume fractions were below the limit of detection for the bookcase and sofa specimens. For the cable specimens, they were fairly repeatable and the yields approached 90 % of the notional yield under well ventilated conditions. Under postflashover conditions, the yields of HCl from the cable specimens were under half the notional yield, indicating that the chlorine was either trapped in the residue or deposited / scavenged somewhere else in the apparatus upstream of the FTIR. Preserving the residue for post-test elemental analysis would have helped address the fate of the unaccounted-for HCl.

- The yields of NO, NO_2, acrolein, and formaldehyde were all below their respective limits of detection.

- Small changes in operating conditions such as air flow, feed rate, and fuel loading did not consistently affect the yields of the measured gases, comparing effects on a single gas between specimen types or between gases for a single specimen type.

- Dicing the specimens had no significant effect, other than allowing low-density materials to be displaced from the specimen boat.

- The sensitivity of the short optical path FTIR cell limited the precision of this examination. However, there was sufficient information to assess the toxicological significance of the variations in the test procedure.

- Calculation of the contributions of the gases to incapacitation of people who might be exposed to these environments showed:
    - For the bookcase material, incapacitation would result from inhalation of CO and CO_2.
    - For the sofa material, incapacitation would result from inhalation of CO and HCN.
    - For the cable material, incapacitation would result from exposure to HCl.

If the CO yield were at the expected postflashover value of 0.2,
    - For the bookcase material, incapacitation would result from inhalation of CO and CO_2.
    - For the sofa material, incapacitation would result from inhalation of CO and HCN.
    - For the cable material, incapacitation would result initially from HCl, and then from CO after a time on the order of 10 minutes, depending on the quantity of material burned.

VII. ACKNOWLEDGMENTS

The authors express their appreciation to Mike Selepak, Kyle Geder, Jonathan Levin, and Adam Schwartz for their assistance in performing the tests and to Jason Averill for the apparatus assembly.

This page intentionally left blank

References

1. Phillips, W.G.B., Beller, D.K., and Fahy, R.F., "Computer Simulation for Fire Protection Engineering," Chapter 5-9 in *SFPE Handbook of Fire Protection Engineering*, 4th Edition, P.J. DiNenno *et al.*, eds., National Fire Protection Association, Quincy, MA, 2008.

2. http://www.nist.gov/el/fire_research/cfast.cfm.

3. Peacock, R.D., Jones, W.W., and Reneke, P.A., "CFAST—Consolidated Model of Fire Growth and Smoke Transport (Version 6) Software Development and Evaluation Guide", NIST Special Publication 1086, National Institute of Standards and Technology, Gaithersburg, MD, 187 pages (2008).

4. Peacock, R.D., Jones, W.W., and Bukowski, R.W., "Verification of a Model of Fire and Smoke Transport," *Fire Safety Journal* **21**, 89-129 (1993).

5. http://fire.nist.gov/fds.

6. "Standard Test Method for Heat and Visible Smoke Release Rates for Materials and Products Using an Oxygen Consumption Calorimeter," ASTM E1354-99, ASTM International, West Conshohocken, PA, 2001.

7. See, *e.g.*, "Standard for the Flammability (Open Flame) of Mattress Sets," 16 CFR Part 1633, *Federal Register* **70** (50), 13471-13523, 2006.

8. http://www.iso.org/iso/iso_catalogue/catalogue_tc/catalogue_tc_browse.htm?commid=50540&published=on&development=on.

9. "Life-threatening Components of Fire – Guidelines for the Estimation of Time to Compromised Tenability in Fires," ISO 13571, International Standards Organization, Geneva, 2012.

10. "Guidelines for Assessing the Fire Threat to People," ISO 19706, International Standards Organization, Geneva, 2011.

11. Gann, R.G., Babrauskas, V., Peacock, R.D., and Hall, Jr., J.R., "Fire Conditions for Smoke Toxicity Measurement," *Fire and Materials* **18**, 193-199 (1994).

12. "Guidance for Comparison of Toxic Gas Data between Different Physical Fire Models and Scales," ISO/DIS 29903, International Standards Organization, Geneva, 2011.

13. "Standard Test Method for Developing Toxic Potency Data for Use in Fire Hazard Modeling," NFPA 269, NFPA International, Quincy, MA, 2007.

14. "Standard Test Method for Measuring Smoke Toxicity for Use in Fire Hazard Analysis," ASTM E1678-10, ASTM International, West Conshohocken, PA, 2010.

15. (a) Gann, R.G., Averill, J.D., Nyden, M.R., Johnsson, E.L., and Peacock, R.D., "Smoke Component Yields from Room-scale Fire Tests," NIST Technical Note 1453, National Institute of Standards and Technology, Gaithersburg, MD, 159 pages (2003).

    (b) Gann, R.G., Averill, J.D., Nyden, M.R., Johnsson, E.L., and Peacock, R.D., "Fire Effluent Component Yields from Room-scale Fire Tests," *Fire and Materials* **34**, 285-314, DOI: 10.1002/fam.1024 (2010).

16. Ohlemiller, T.J., and Villa, K., "Furniture Flammability: An Investigation of the California Bulletin 133 Test. Part II: Characterization of the Ignition Source and a Comparable Gas Burner," NISTIR 4348, National Institute of Standards and Technology, Gaithersburg, MD, 1990.

17. Gann, R.G., Marsh, N.D., Averill, J.A., and Nyden, M.R., "Smoke Component Yields from Bench-scale Fire Tests: 1. NFPA 269/ASTM E 1678," NIST Technical Note 1760, National Institute of Standards and Technology, Gaithersburg, MD, 57 pages (2013).

18. Kaplan, H.L., Grand, A.F., and Hartzell, G.E., *Combustion Toxicology: Principles and Test Methods*, Technology Publishing Co., Lancaster, PA, 1983.

19. "Guidance for Evaluating the Validity of Physical Fire Models for Obtaining Fire Effluent Toxicity Data for Use in Fire Hazard and Risk Assessment – Part 2: Evaluation of Individual Physical Fire Models," ISO/TR 16312-2, International Standards Organization, Geneva, 2007.

20. "Controlled Equivalence Ratio Method for the Determination of the Hazardous Components of Fire Effluent," ISO/TS 19700, International Standards Organization, Geneva, 2006.

21. "Reaction-to-fire Tests – Heat Release, Smoke Production, and Mass Loss Rates – Part 1: Heat Release (Cone Calorimeter Method)," ISO 5660-1, International Standards Organization, Geneva, 2002.

22. "Plastics — Smoke Generation — Part 2: Determination of Optical Density by a Single-chamber Test, ISO 5659-2, International Standards Organization, Geneva, 1994.

23. Blomqvist, P., Hertzberg, T., Tuovinen, H., Arrhenius, K., and Rosell, L., "Detailed determination of smoke gas contents using a small-scale controlled equivalence ratio tube furnace method," *Fire and Materials* **31**, 495-521, doi:10.1002/fam.946 (2007).

24. Stec, A.A., Hull, T.R. & Lebek, K. "Characterisation of the steady state tube furnace (ISO TS 19700) for fire toxicity assessment," *Polymer Degradation and Stability* **93**, 2058-2065, doi:10.1016/j.polymdegradstab.2008.02.020 (2008).

25. Stec, A.A., Hull, T.R., Purser, J.A. & Purser, D.A. "Comparison of toxic product yields from bench-scale to ISO room," *Fire Safety Journal* **44**, 62-70, doi:10.1016/j.firesaf.2008.03.005 (2009).

26. Rhodes, J., Smith, C. & Stec, A.A. "Characterisation of soot particulates from fire retarded and nanocomposite materials, and their toxicological impact," *Polymer Degradation and Stability* **96**, 277-284, doi:10.1016/j.polymdegradstab.2010.07.002 (2011).

27. Stec, A.A. & Rhodes, J. "Smoke and hydrocarbon yields from fire retarded polymer nanocomposites," *Polymer Degradation and Stability* **96**, 295-300, doi:10.1016/j.polymdegradstab.2010.03.032 (2011).

28. Haaland, D.M.; Easterling, R.G.; and Vopicka, D.A., "Multivariate Least-Squares Methods Applied to the Quantitative Spectral Analysis of Multicomponent Samples" *Applied Spectroscopy* **39**, 73-84. doi:10.1366/0003702854249376 (1985).

29. *Gas Phase Infrared Spectral Standards, Revision B*, Midac Corp.; Irvine, CA (1999).

30. Speitel, L.C., "Fourier Transform Infrared Analysis of Combustion Gases," Federal Aviation Administration Report DOT/FAA/AR-01/88, 2001.

31. Pitts, W.M., Hasapis, G., and Macatangga, P., "Fire Spread and Growth on Flexible Polyurethane Foam," 2009 Fall Technical Meeting of the Eastern States Section of the Combustion Institute, University of Maryland, College Park, MD, October 18-21, 2009.

32. Pitts, W.M., "The Global Equivalence Ratio Concept and the Formations Mechanisms of Carbon Monoxide in Fires," *Progress in Energy and Combustion Science* **21**, 197-237, (1995).

33. "Smoke Gas Analysis by Fourier Transform Infrared Analysis: The SAFIR Project," VTT Research Note, Technical Research Centre of Finland, 81 Pages, 1999.

34. "Toxicity testing of fire effluents -- Guidance for analysis of gases and vapours in fire effluents using FTIR gas analysis," ISO 19702, International Standards Organization, Geneva, 2006.

35. Kaplan, H.L., Grand, A.F., Switzer, W.G., Mitchell, D.S., Rogers, W.R., and Hartzell, G.E., "Effects of Combustion Gases on Escape Performance of the Baboon and the Rat," *Journal of Fire Sciences* **3**, 228-244, (1985).